AF553931

# INDUSTRIAL BIOTECHNOLOGY

*By*

**Dr. P.R. Yadav**

*Lecturer*

*Department of Zoology*

*D.A.V. College*

*Muzaffarnagar (U.P.)*

*&*

**Dr. Rajiv Tyagi**

*Department of Zoology*

*M.M. College*

*Modi Nagar (U.P.)*

DPH

DISCOVERY PUBLISHING HOUSE

NEW DELHI-110002

First Published – 2005

Reprinted – 2025

ISBN: 978-81-8356-038-2



**Industrial Biotechnology**

*Published by:*

**DISCOVERY PUBLISHING HOUSE**

4383/4B, Ansari Road, Darya Ganj
New Delhi-110 002 (India)
*Phone*: +91-11-23279245; 23253475; 43596065
*Mobile*: +91 9811179893 / +91 9871656464
*E-mail*: discoverybooksindia@gmail.com
orderdphbooks@gmail.com
*web*: www.discoverypublishinggroup.com

*Printed at:*
Infinity Imaging Systems
Delhi

# Preface

The present title aims to attract the interest of students in industrial biotechnology at the undergraduate and post-graduate level as well as that of everybody interested in basic research in molecular genetics, bio-organic chemistry, biochemistry, agrobiotechnology, pharmaceutical science and medicine. Industrial biotechnology by its iterative rational design is a powerful tool to both test general theories as well as to develop more useful techniques to be used in biotechnological processes or products.

Text book can not be written without the support and professional contributions of many peoples. This book is no exception. The authors are grateful to those teachers and colleagues whose stimulating discussions have clarified certain vague points for him, but all errors, omissions, and collcctions needed and solely their responsibility.

In the preparation of this book large number of books and research papers have been consulted. So no authenticity is claimed.

The authors tried hard to be accurate and upto date in statement and realises the impossibility of completely avoiding errors therefore, the authors will greatly appreciate having his attention called to any questionable statements.

The authors express their gratitude to Mr. Wasan and staff of M/s Discovery Publishing House for their whole hearted co-operation in the publication of this book.

**Authors**

# CONTENTS

1. **Introduction** 1—32

Cultivation of Microbes, Developing an Industrial Process, Types of Fermentation Processes, Industrial Ethyl Alcohol Manufacture, Alcoholic Beverages, Distilled Alcohol as Beverage and Fuel, Production of Butanol, Production of Vinegar, Foods From Waste, Amino Acid Production, Microbial Enzymes in Industry, Vitamins and Amino Acids, Antibiotics and Steroids, Microbiological Assay, Industrial Spoilage, Prevention of Spoilage.

2. **Biogas Generation and Comparisons** 33—46

Description of Systems, Suspended Growth Systems, Supported Growth Systems, Applications, Treatment Plant Sludges, Feed Lot Wastes, Rural Installations, Food Processing Wastes, Pretreatment.

3. **Ethanol Production** 47—57

Feedstocks, Sugar and Starch, Lignocellulosics, Acid vs. Enzymatic Hydrolysis of Lignocellulosics, Acid Hydrolysis of Lignocellulosics, Enzymatic Hydrolysis of Lignocellulosics, Major Component Steps in an Enzyme Based Biomass-to-Ethanol Process, Pretreatment, Fractionation, Hydrolysis, Fermentation, Hexose Fermentation, Pentose Fermentation, Ethanol Recovery, Current and Future Status of Enzymatic Hydrolysis.

4. **Food Preservation** 58—138

Benzoic Acid and its Salts, NaCl and Sugars, Sulfur Dioxide and Sulfites, Propionates, Nitrites and Nitrates, Organisms Affected, Perigo Factor, Interaction with Cure Ingredients and Other Factors, Nitrosamines, Nitrite-sorbate and other Nitrite Combinations, Mode of Action, Summary of Nitrite Effects, Flavouring Agents, Indirect

Antimicrobials, Antioxidants, Medium-chain Fatty Acids and Esters, Acetic and Lactic Acids, Antibiotics, Spices and Essential Oils, Nisin, Natamycin, Subtilin, Tetracylines, Tylosin, Ethylene and Propylene Oxides, Antifungal Agents for Fruits, Miscellaneous Chemical Preservatives, With Low Temperature, Temperature Growth Minima of Food-borne Microorganisms, Preparation of Food for Freezing, Freezing of Food and Freezing Effects, Storage Stability of Frozen Foods, Effect of Freezing upon Microorganisms, Effect of Thawing, High Temperature, Factors that Affect Heat Resistance in Microorganisms, Water, Fat, Salts, Carbohydrates, pH, Proteins and other Substances, Numbers of Organisms, Growth Temperature, Inhibitory Compounds, Time and Temperature, Relative Heat Resistance of Microorganisms, Thermal Destruction of Microorganisms, Thermal Dath Time (TDT), D. Value, Z Value, F Value, Thermal Death Time Curve, 12-D Concept, Aspetic Packaging, By Drying up the Food, Preparation and Drying of Low-Moisture Foods, Effect of Drying Upon Microorganisms, Storage Stability of Dried Foods, Intermediate-moisture Foods (IMF), Preparation of IMF, Microbial Aspects of IMF, Storage stability of IMF, Preservation by Radiation, Ultraviolet Irradiation, Germicidal Lamps, Factors Influencing Effectiveness, Effects on Humans and Animals, Action on Microorganisms, Applications in the Food Industry, Ionizing Radiations, Kinds of Ionizing Radiations, Definition of Terms, X-Rays, Gamma Rays and Cathode Rays, Effects on Microorganisms, Effects on Foods, Applications, Microwave Processing.

5. **Yeast in Industries** **139—157**

Transformation, Procedures, Genetic Markers for Yeast Transformation, Yeast Cloning Vectors, Integrating Vectors, Extrachromosomally-Replicating Vectors, DNA Cloning in Yeast Plasmids, DNA Cloning in Yeast Cosmids, Fusion Plasmids, Vectors for High Level Gene Expression, Industrial Applications.

6. **Large-scale Fermentation** **158—173**

Design and Operation of Contained Fermentors, Primary Containment Features, Ancillary Containment Features, Computer System, Cell Growth and Product Regulation, Control of Cell Growth, Product Biosynthesis and Accumulation.

7. **High Fructose Corn Syrup** **174—186**

Historical Perspective, Development of Key Enzymes, High Fructose Corn Syrup Products, Modern Process Technology,

Manufacture of Dextrose from Starch, Preparation of Dextrose Feedstock for Isomerization, Isomerization Process, Secondary Refining of Isomerized Feedstock, Manufacture of Enriched Fructose Products, Marketing and Economic Considerations.

**8. Fermentation Reactions** 187—203

Metabolic Groups and Pathways, Two- and Three-phase Models, Interspecies Hydrogen-Transfer, Hydrolysis and Initial Reactions of Anaerobic Digestion, Hydrogen Concentration and Effects on Metabolic Pathways, Factors Affecting Rate and Extent of Methanogenesis, Rate-Limiting Reactions, Temperature, pH, Stoichiometry, Nutrition and Inhibition.

**9. Control of Forest Pests** 204—218

Pests of Forests, Vegetation Management, Plant Pathogens, Nematodes, Phytoalexins, Viral Pathogens of Plants, Insect Pests, Viral Pathogens of Insects, Modified NPVs for Insect Control, Alternative ways of Improving Baculoviruses, Other Applications of Biotechnology for Pest Control, Conclusions.

**10. Drugs from Plants** 219—236

Biotic Resource, Role of Plants in Drug Discovery and Development, Progress in Plant Drug Research, Some Significant Plant Drugs, Drugs for Heart Diseases, Local Anaesthetics, Analgesics, Antimuscarinics, Miotics, Muscle Relaxants, Bronchodilators, Antineoplastic Agents, Antiprotozoals, Other Miscellaneous Drugs from Plants, Recently Discovered Plant-based Drugs, Taxol and Camptothecins, Artemisinin, Summary and Conclusions.

**11. Packaging Food** 237—271

Sanitary Cans, Glass Containers, Paper Containers, Requirements of a Container, Basic Factors, Consumer Acceptance Factors, Rating the Container, Weighted Values of Factors, Discussion of Grading Factors, Applied to Different Types of Containers, Age of Productive Packaging, Water Vapour Transmission, Bursting Strength, Tensile Strength and Elongation, Internal Tearing Resistance, Grease Resistance, Gas Permeability.

**12. Recombinant DNA Vaccines** 272—285

Rationale, Methods, Preparation of DNA, DNA Cloning Vectors, Identification of Recombinants, Development of an r-DNA Vaccine to Prevent ETEC Induced Diarrhea, Rationale for Development, Cloning of E. coli Pilus Genes, Subcloning, Final Remarks.

**Index** 286—290

# 1

# INTRODUCTION

Since prehistoric times, humans have taken advantage of the beneficial activities of microbes. However, it has only been within the past forty years that these activities have been harnessed for the large-scale production of microbial cells or their products. Microbiologist, engineers, and businesses have come together to develop the field of industrial microbiology. By controlling the activities of certain microbes, they have made possible the simpler, more economical production of large quantities of useful products such as enzymes, amino acids, vitamins, antibiotics, organic acids and alcohol.

Yeast, bacteria, and molds are grown in containers with capacities are large as 50,000 gallons. For example, yeast cells are grown for use in ironized yeast tablets for human use, and in even larger quantities as a supplement to farm-animal feed. It is not an easy job to handle large quantities of growing microbes. Expensive equipment and well trained personnel are essential to the smooth operation of an industrial process such as antibiotic or beer production. Industrial microbiologists must closely monitor the chemical activities of the microbes and be prepared to change or stop the organism's activities to prevent product loss. Careful attention must also be paid to ensure that only one kind of microbe is being grown. Successful control of these factors has resulted in the development of multimillion-dollar microbial industries.

## CULTIVATION OF MICROBES

Their small size, short generation time and rapid metabolic rate make microbes highly productive. Controlling microbial metabolic acclivities has enabled industrial microbiologists to develop three fundamental industrial processes. The first is the *cultivation* and *harvesting* of cells that may be used as food, food supplements, or inocula for other industrial processes. An example of this type of industrial process is the production of activated dry yeast and ironized yeast tablets. The second type of process utilizes select microbes as "factories" that enzymatically convert a particular substrate into a desired product, a process known as *bioconversion*. Steroids known as estrogens and progesterones used in birth control pills are produced by the action of microbes on specific substrates added to culture medium. The third industrial process utilizes microbes for their ability to generate large quantities of a specific *metabolic byproduct*. Alcohol, enzymes, and many types of organic acids are commonly produced by the microbial fermentation of inexpensive and readily available substrate material.

The industrial cultivation of microbes is carried out in large tanks known as fermenters. These units range in capacity from 5000 gallons (about 20,000 liters) to 50,000 gallons (about 200,000 liters). Some of the more basic and inexpensive substrates used as culture media are molasses (the thick syrup made from boiling sweet vegetables, sugar cane, or fruits), soybean meal, and malted grain. Fro specific bioconversion processes, more expensive synthetic media may have to the prepared from large quantities of purified ingredients. The fermenter is inoculated with a volume of culture equal to 5 to 10 per cent of the tank's volume. This means that a 15,000-gallon tank requires about 1500 gallons of pure culture to start the fermentation. This inoculum is built-up from a pure agar slant culture of microbes maintained in the laboratory, or it may be bought from a microbiological supply house. The initial culture is inoculated into ever increasing amounts of media to reach the final volume necessary for the processes. This repeated subculturing requires a great deal of space and close control to assure that contamination does not occur. In order to reduce these problems, lyophilized cultures can be purchased from supplies who have concentrated the required volume into a more manageable container. Once the substrate has been inoculated, all the conditions necessary for growth and reproduction must be accurately maintained. The optimum temperature, pH, purity, salinity, and cell density must be checked regularly. Since many industrial processes are aerobic, an adequate air supply must be made available to these fermenting microbes. This is done by injecting sterilized air into the tank through inlet valves at the bottom of the fermenter and continuously stirring the culture with mechanically operated paddles installed inside. The

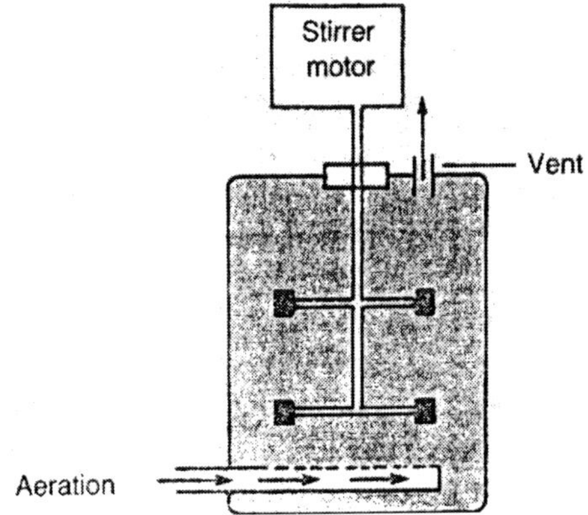


*Fig. 1.1. A typical batch fermentation.*

rapid bubbling and stirring causes a layer of foam to form on the surface that is controlled by the addition of antifoaming chemicals.

The two different methods that have been developed for industrial fermentations are *batch* and *continuous* culturing. Batch cultures are made by adding the microbes to a limited quantity of medium and allowing the fermentation to run to completion without the addition of fresh nutrients or removing by-products. Cheese is made by batch culturing. Once inoculated into the fermenter, the microbes reproduce and increase in number according to the population growth curve. The maintenance of cultures in the exponential (log) phase of growth is known as steadystate, or balanced, growth and allows microbes to be cultured on a continuous basis. Continuous cultures are maintained in

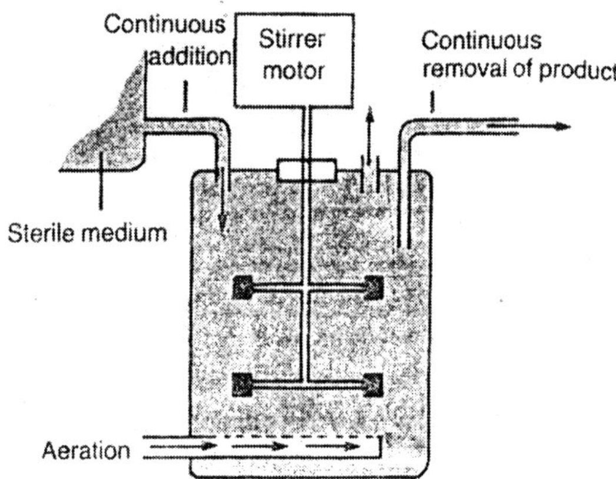


*Fig. 1.2. Continuous culture unit.*

an apparatus called a *chemostat*. Many antibiotics, organic acids, and steroids are produced by microbes in continuous culture. The log phase of growth is maintained by continuously adding fresh, sterile medium to the *fermenter* at a rate equal to the removal of culture and products. The exponential growth rates regulated by maintaining the precise level of a limiting factor in the medium. This practice keeps the cells from shifting into a lag, or stationary, phase. Continuous cultures have an advantage over batch cultures, since the product can be constantly manufactured in large amounts. However, minor chemical or physical changes in the medium or spontaneous mutations within the microbial population prevent this method from being carried out longer than a few days.

Once the fermentation is complete, the product must be separated from the culture medium and purified. Passing the culture through a filter separates the medium from the microbes. If cells are the desired product, they are washed and concentrated by centrifugation. If the product is dissolved in the medium, it is separated by mixing with a solvent which selectively extracts the product from unwanted compounds. Depending on the products type, it is then separated from the solvent and crystallized in pure form. In some cases, the chemical may not be suitable in a crystalline form and must be redissolved for chemical modification. Members of the penicillin family of antibiotics result from chemical modifications of the basic penicillin molecule crystallized from such a process.

With an increased understanding of organic chemistry, it has become possible to selectively modify many microbially produced compounds to the exact chemical structure needed for a particular purpose. However, successes in the chemical industry have not always benefited industrial microbiologists. Many of the organic compounds through to be produced only by microbes are now being manufactured in the chemistry lab. These products are manufactured in larger quantities from low-cost starting materials in less sophisticated equipment, enabling them to be marketed at a lower price. To overcome this problem, industrial microbiologists constantly search for new microbes or techniques to produce complex, commercially important compounds. One method used to increase production and reduce costs has been the development of productive mutations. *Productive mutants* are microbes that have been produced by natural selection following exposure to mutagenic levels of radiation. This technique was successfully used to increase penicillin production from the fungus *Penicillium chrysogenum* during World War II. Spores of this mold were isolated and irradiated with high dosages of ultraviolet light which caused mutations in the genes. When some of these spores were germinated, the new mycelia were found to produce larger amounts of the antibiotic than the original.

**Developing an Industrial Process**

In developing an industrial process based upon the action of micro-organisms many details must be given consideration. Important among these are the type of culture needed, cultural conditions and the adaptability of the organism to large-scale production.

***Purity and nature of cultures***

It must be ascertained whether absolutely pure cultures must be used, or whether the mere predominance of one organism is sufficient. This may be a deciding factor, as the cost of preparing and maintaining pure cultures and sterile apparatus throughout a process is relatively high.

***Cultural conditions***

The organism must be able to grow well in the medium to be used and under the conditions of the process. This necessities exact studies and careful control of *optimum* conditions of aerobiosis or anaerobiosis, temperature, nutrition and pH. Appropriate adjustments of the process and apparatus must be made to provide the necessary conditions.

***Productive mutants***

The organisms selected must be such as will produce the desired substance(*s*) or results in the medium under the conditions furnished, in amounts sufficient to yield a profit. Some firms have "per" strains of micro–organisms that excel in producing certain products, such as butyl alcohol, certain antibiotics or itaconic acid, that they have "developed" (selected mutants) for these purposes. It has been found possible to induce industrially valuable mutations in micro–organisms by ultraviolet radiations. Where sexual processes are known to occur, the breeding of yeasts and molds for similar purposes is analogous to breeding of farm animals for special purposes.

***Medium or raw material***

The substrate or medium should support luxuriant growth of the organism to be used and it must be available constantly at costs compatible with profit. Expensive handling machinery may be needed for some substrates.

An important item is the possible necessity of a preliminary treatment such as liming of very acid yeast slops, distillery wastes, molasses and whey. Some substrates, such as sawdust or fiber, may need preliminary "digestion" with hot acid or alkali to hydrolyze them to fermentable substances. This all adds to the expense and time.

***Nature of the process***

The more complicated and exacting the system of cultural details and preliminary heatings, dilutings and digestions, as well as the type of machinery (cracking stills, tanks, pumps) to handle the end-and by-products and the final wastes, the greater will be the cost and therefore the less the commercial practicability of any process. Any time-consuming aging or ripening process eats into profits. Sometimes, very desirable end or by-products may be found in commercial fermentations, yet the cost of their recovery may be prohibitive.

***Preliminary experimentation***

The microbiologist working with 10-ml. test-tube cultures may find many valuable things. When attempts are made to reproduce the test-tube experiments on a 100,000 gallon factory scale, however, the laboratory discoveries often fail to yield the promised result. Any process developed in the experimental laboratory must next prove its worth in the factory. A small-sized model, or pilot plant, is usually tried after the preliminary laboratory work. All may depend on such a seemingly far-removed detail as international relations. These may affect the cost of importation of some raw product essential to the process under investigation. Then the industrialist turns to home resources, goes to Washington, or *employs a resourceful microbiologist*!

The whole matter is a complex of micro-biology, chemistry, engineering and economics. Only the microbiology can be discussed here. Many chemical and microbiological processes in use at present are patented and secret, and specific strains of bacteria, yeasts and molds, which are zealously guarded, are often carefully developed in the laboratories of manufacturing concerns. As a result of continuous and intensive industrial research, methods change or are superseded frequently.

## Types of Fermentation Processes

Industrial fermentation processes may be divided into two main types: (1) batch fermentation and (2) continuous process. There are various combinations and modifications of these.

***Batch fermentations***

A tank or fermentor is filled with the prepared *mash* (material to be fermented, e.g., diluted molasses, comminuted potatoes, digested corn cobs). The proper adjustments of pH, temperature, nutritive supplements and so on are made. In a pure-culture process, the mash is steam-sterilized, the entire fermentation tank sometimes being the autoclave. The inoculum, a *pure culture*, is added from a separate pure-culture apparatus. The fermentation proceeds. Some pressure may be maintained within the tank to prevent inward leakage of contamination and sometimes to maintain increased tension of special gases. After the proper time, the contents of the fermentor are drawn off for further processing, the fermentor is cleaned, and the process begins over again. Each fermentations is a discontinuous process divided into *batches*.

***Continuous-growth process***

In continuous-growth processes, the substrate is fed into a container continuously at a fixed rate. The cells grow (or enzymes act) continuously as the material passes through the apparatus. The organisms and process are said to be in a *Steady state* or condition of *homeostasts*. The product or fully fermented mash is drawn off continuously. The engineering arrangements may be complex, permitting aeration, cooling or heating adjustment of pH or addition of nutrients continuously during the process. There must also be means of controlling rate of growth, phase of growth curve, and removal of dead organisms. The culture must remain pure and must not undergo any variation.

One may conceive of such a process as taking place in a long pipe (actually it may be a roating conical tank or series of connected tanks). At one end the prepared mash enters. It at once encounters the growing organisms. These act on the substrate as it flows through the system. At the stage at which the valuable product of the fermentation is at its maximum concentration, the fluid is drawn into receiving vessels for further processing (e.g., distillation). The animal alimentary tract may be thought of as a natural, continuous-growth process.

***Submerged aerobic cultures***

Many industrial processes, casually called "fermentations," are carried on by strictly aerobic micro-organisms: for example, production of penicillin by *Penicillium notatum*, a strictly aerobic mold. In older aerobic processes it was necessary to furnish large surfaces of culture media exposed to air. The limitations of space, difficulties from contamination, and expense of hand labour can well be imagined, though little expense for power equipment was necessary. Now it is common commercial practice to carry on such "fermentations" in closed tanks with *submerged cultures*. Aerobic conditions are maintained by constant agitation of the contents of the tank with an *impeller* and constant *aeration* by forcing sterilized air through a porous *diffuser*. The flow of air through the tank removes gases such as ammonia and carbon dioxide. In each sort of process very careful adjustments of O-R potentials, mechanical agitation, ratio of dissolved oxygen to other ingredients in the medium, and pH are necessary. This is one of the many fascinating and potentially very lucrative fields for research in industrial microbiology.

**Industrial Ethyl Alcohol Manufacture**

Much industrial ethyl alcohol is now made from by-products of *cracking* petroleum to make gasoline. However, the manufacture of ethyl alcohol from fermentation by yeasts is still an important industry. It serves to illustrate industrial fermentation processes in general. Crude molasses is often used as mash. It generally requires only to be diluted and the pH adjusted (usually with sulphuric acid) to 4.5. This pH is favourable to the yeast and unfavourable to many bacteria. A source of nitrogen such a ammonium sulphate or ammonium phosphate is usually added. The final solution is a richly nutrient carbohydrate culture medium of *mash*.

This is rather heavily inoculated with an aciduric and alcohol-resistant strain of yeast, the variety depending on the conditions under with the fermentations is to proceed and the exact end products desired. A good strain of *Saccharomyces cerevisiae* is commonly used. The inoculum comes from a large tank of carefully maintained pure culture, previously inoculated from a smaller seed tank, and the latter from a flask or tube of culture in the laboratory. At present stainless steel continuous-culture apparatus is available for maintaining constantly large amounts (many gallons) of *pure* cultures of inoculum.

The maintenance of purity of the inoculum is a responsibility of the microbiologist, and woe betide him if some sporeformer, *Lactobacillus*, wild yeast or bacteriophage gets in and ruins 100,000 gallons of mash! The mash and all of the machinery are generally sterilized before the inoculation and then cooled. The microbiologist is kept busy at every stage of the process, making cultural and microscopic examinations of the water, mash and apparatus to detect and eliminate contamination.

In the batch process, much used for this purpose, fermentation in enormous tanks is allowed to continue for about 48 hours at a carefully controlled temperature of about 25°C until the yeast stops growing because of the concentration of alcohol and other products. Aeration

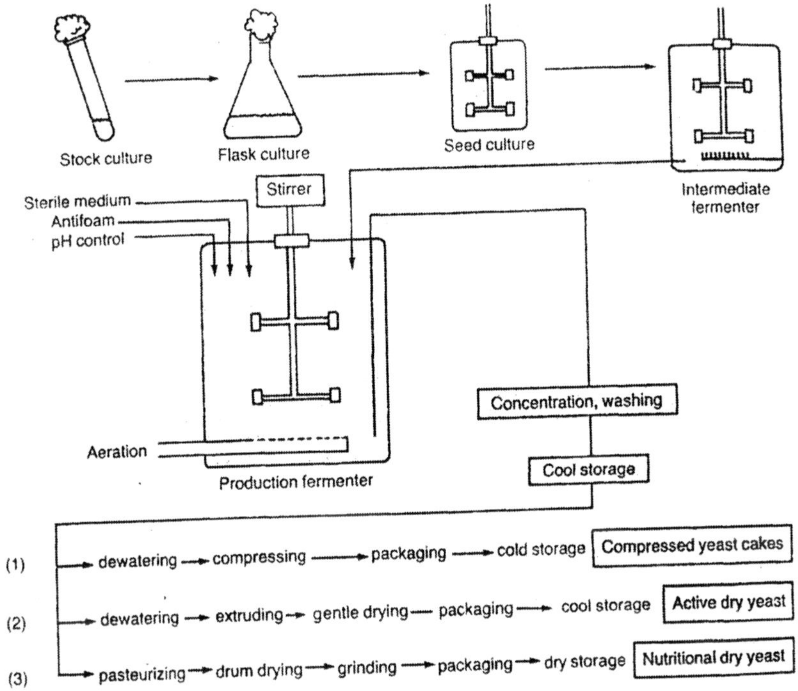

*Fig. 1.3. The industrial production of yeast, Saccharomyces cerevisiae.*

with filtered air is used at first to promote rapid growth, but anaerobiosis is soon established to promote fermentation and alcohol accumulation, and prevent its oxidation to carbon dioxide and water.

The fermented mash contains the crude alcohol or *high wine*, as it is called. This is usually a mixture of ethyl alcohol and a small amount of glycerol with *fusel oil*. The last contains amyl, isoamyl, propyl, butyl and other alcohols with acetic, butyric and other acids, as well as various esters. The high wine is driven off from the mash or *beer* by heart, and further purified by fractional distillation, which is a problem in chemical engineering.

The Chemical reactions involved in the fermentation are complex; the principal stages follow the Embden-Meyerhof scheme. The overall reaction in the production of alcohol from glucose is:

$$C_6H_{12}O_6 \xrightarrow{\text{yeast}} 2C_2H_3OH + 2CO_2.$$

The chief constituents of fusel oil are probably derived from the action of the yeasts on amino acids in the mash. The large amounts of carbon dioxide evolved are purified and compressed in tanks or made into solid carbon dioxide. Part of this may be used for cooling the fermentation vats.

## Alcoholic Beverages

### *Wine*

Alcoholic beverages can be made by fermenting almost any grains, fruits, or vegetables that contain sugar. It was not uncommon for the early colonists in the United States to ferment pumpkins, potatoes, Indian corn, or carrots to make a beer. They fermented and distilled honey to make a rum substitute known as "old methaglin." Apple cider (unfiltered and non-pasteurized apple juice) was fermented to "hard cider." Hard cider was sometimes placed in wooden barrels and frozen in a lake during the winter to separate the water and concentrate the alcohol. This "Cold distilled" beverage was known as applejack and was very potent. Some prepared a mixture of applejack and gin in a drink known as "strip-and-go-naked" because of the strange behaviour it caused in those who drank only a few mugfuls. Other fermented fruits included elderberries, plums, peaches, currents, raspberries, and grapes. Today grapes are the primary fruit used in the production of wine. *Wine* is the fermented juice of grapes and other fruits with an alcohol concentration of between 9 and 21 per cent. Table or dinner wines contain no more than 14 per cent alcohol, while dessert and appetizer wines may contain up to 21 per cent alcohol. *Beer* is the fermented extract of grains such as barley or rice with an alcohol concentration of between 3 and 6 per cent. *Packaged liquors* are distilled fermentation products that have alcohol concentration as high as 85 per cent.

Grape wine may be made from any of 5000 varieties of the plant *Vitis vinifera* or 2000 varieties of other *Vitis* species. Each has a characteristic chemical contents that imparts a unique flavour, aroma (bouquet) and colour to the wine made from each variety. Because the chemical content of the grape is greatly affected by the soil and climate in which it is grown, grapes used for making good wines are only grown within certain wine-growing regions around the world.

Grapes contain a variety of sugars but the two most important are glucose and fructose. Glucose is fermented by yeast to carbon dioxide and alcohol, and the fructose gives a sweet taste to the wine. Since fructose tastes twice as sweet as glucose, grapes with a high percentage of this sugar are excellent for the production of sweet white wines. Also found are organic acids, free amino acids, proteins, pigments, and various minerals. The particular balance of these nutrients and

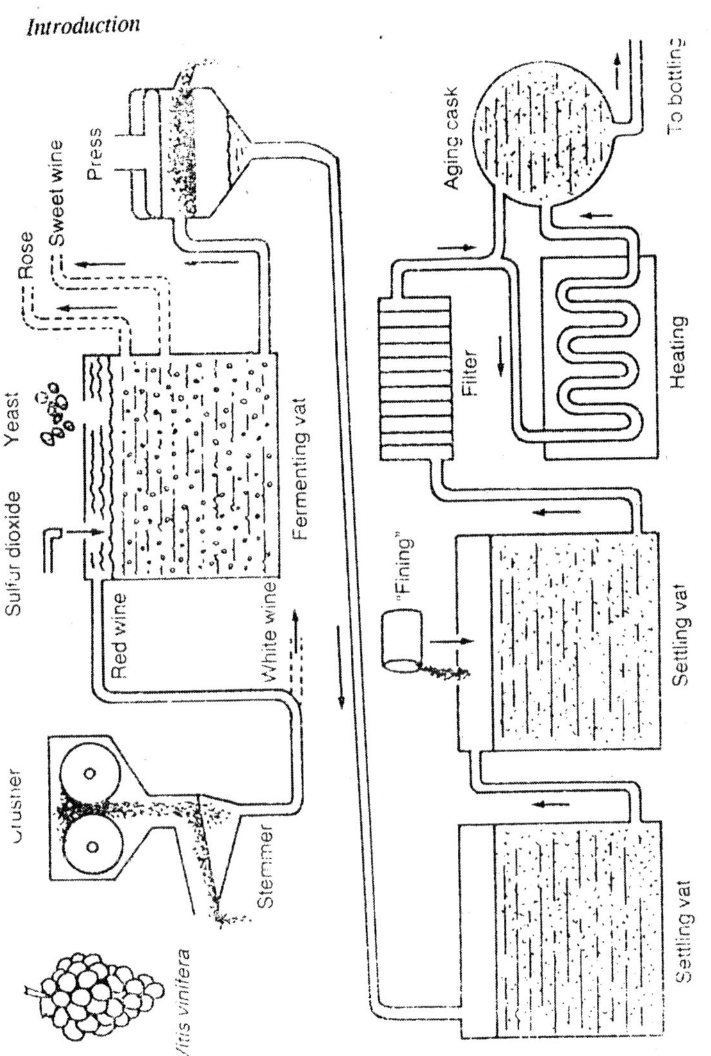


*Fig. 1.4. This is only one method among many that might be used commercially to prepare wine.*

the wine-making process affects the characteristics and quality of a wine. Grapes are picked from the vine at a time when they are most likely to have the desired balance of sugar and other compounds. They are quickly taken to the winery where they are mechanically crushed.

The stems are separated from this mixture of pulp, skins, and seeds called "must". The skins are also removed if white wine is to be made. Red and rose wines get their colour and flavour from pigments and tannic acid in the skins. "Must" is pumped to the fermenting vat made of hardwood or glass-lined concrete. Once inside the vat, sulphur dioxide or sulphurous acid is added to inhibit the growth of wild, undesirable strains of yeast that occur naturally on the grapes. It also interferes with natural enzymes that "brown" the wine, destroys grape skin cells aiding in the release of red pigments. Acidifies the must, and aids in clarifying the wine after fermentation. Pure cultures of *Saccharomyces cerevisiae* var, *ellipsoideus* are unaffected by the sulphur dioxide and are added to the vat to ferment the must for a few days to a few weeks. Heat generated by the yeast is carried away by cooling coils in order to maintain a fermentation temperature below 85°F for red wines and below 60°F for white wines. An increasingly popular wine in the United States is rose. The production of rose wines begins as with any other red wine, but the skins are removed early in the initial fermentation to prevent the development of a dark red colour and heavier tannic acid concentration.

When this portion of the wine-making process is complete, the wine is pumped to a wine press to remove the pieces of skins, seeds, and other solids. The wine is present to a settling vat where the fermentation is allowed to continue, while smaller solids and yeast cells settle out. During this time, *lactobacillus* species convert strong malic acid to weak acids, thus reducing the sharp flavour of the wine. When the wine is clear and has matured, it is "racked," or drawn off, to another settling tank. There the filtering process further clarifies the wine and reduces its oxygen content. Wines that lose their carbon dioxide during this process are called *still wines*. In many large wineries, clarification is added by the addition of "fining" agents such as clay, gelatin, or egg white which precipitates extremely small particles in the wine. After clarification and fining, it is pumped into oak barrels or casks for aging.

During the aging process, chemical reactions occur which develop the wine's flavour, aroma, and colour. The yeasty flavour is lost and red wines develop their typical rich red colour, while white wines become amber. The aging process usually requires about two years for red wines and varies from several months to about two years for white wines. The decision to bottle is made by the wine master and is very crucial in the production of a fine wine. However, the aging process does not stop after bottling. Red wines continue to develop and mature for years after bottling. Most wine masters and connoisseurs feel that red wines are their best five to ten years after bottling and white wines reach their peak after about two years.

*Sparkling wines* and champagnes are made in a unique way to capture and retain the natural effervescence of the carbon dioxide released during fermentation. The sparkling wine *Champagne* is only made in the Champagne region of France, while the uncapitalized name champagne is used as a generic name in the United States to describe any wine that contains an excess carbon dioxide level. When Champagne is made, the must is fermented and fined as with all wines. However a second fomentation is initiated after the wine is bottled by the addition of more yeast and sugar. These specially designed bottles are wired shut to prevent the gas from escaping and allowed to ferment neck-down on a special rack for a few weeks to several months—a process called tirade.

When the fermentation is complete, dead yeast, tartar, and other compounds settle on the inside of the bottle. This sediment must be removed without disturbing the carbonation or quality of the wine. This is done by "riddling," a special bottle-turning done by hand that causes the sediment to drop into the neck of the bottle. A talented riddler can riddle, or turn, about 30,000 bottles of Champagne a day. Each bottle must be turned once a day for about a month to sediment all the yeast into the neck. The bottles are then cooled almost to freezing, so that they can be turned upright and the corks removed along with the frozen sediment. It is at this time that sugar syrup is added to give the Campagne a more appealing taste. Four different types of Champagne are produced and vary according to the percentage of sugar added at this time. Brut is the driest (0.5-1.5 per cent sugar), sec has 2.5-4.5 percent sugar, demi-sec is more sweet (5 per cent), and doux the "sweetest" with 10 per cent added sugar. If riddling and tirage have been done well, the product is a clear, sparkling wine of high quality.

***Beer***

Another popular alcoholic beverage, beer, is believed to have been first produced by the Egyptians as the direct product of the fermentation of barley. As with wine the production of a fine beer requires the skillful blending of science and art. Several different types of beer are made throughout the world and vary in their colour, aroma, flavour, and alcohol content.

The brewing of beer begins with the selection of high quality barley and a process known as malting. *Malting* involves steeping (soaking) the grain in water of 65°C to initiate germination, or sprouting. During seed germination, natural enzymes convert the stored food

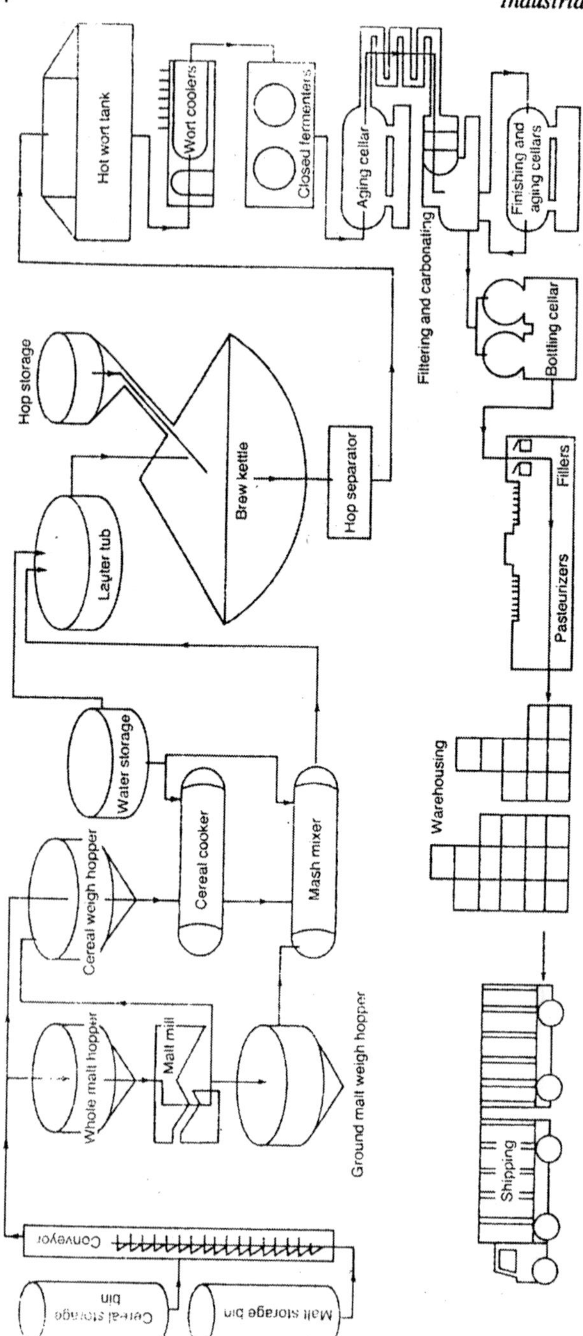


*Fig. 1.5. The brewing process.*

material in the seed (endosperm) from starch to glucose. This increases the percentage of glucose available for yeast fermentation and reduces the amount of starch to a level that prevents the development of fusel oil during fermentation. The malted grain is then dried in kilns or rotatining drums at from 75° to 100°C. The actual temperature used for drying depends on the kind of malt used in the preparation of the brew. Lower temperatures are used for light colour malt and higher temperatures are used for dark malt. Care must also be taken to preserve the barley enzymes (amylase) needed to convert additional starch to sugar during the *mashing* process. In this process, the malt is cleaned, ground, weighed, and mixed with other cereal grains such as corn meal, soybean meal, corn grits, or corm flakes. These *adjuvants* help balance the protein and carbohydrate content to ensure a more flavourful product. The mixture is sent to a lauter tub, where it is purged with hot water to extract the nutrients from the mixed grains. The leftover of "spent" grain is dried and sold for used as animal feed. The protein content of this grain is very high, thus it is an excellent supplement to cattle and hog feed. The fluid extract, known as *wort* (pronounced *"wert"*), has the colour of beer and is pumped to the brew kettle where it is cooked for 1½ to 2½ hours. While in the brew kettle, flowers of the female hop plant are added to give beer its characteristic aroma and inhibit the growth of contaminating bacteria that could later interfere with fermentation or cause defects. The most important contaminating bacteria are *Lactobacillus pastorianus*, *Pediococcus cerevisiae*, *Flavobacterium proteus*, and *E. coli*.

After cooking in the brew kettle, the wort is passed through a hop separator that not only removes the hops but also extracts undesirable resins and residues. The clarified wort is then cooled and pumped into a closed fermenter where an active culture of *Saccharomyces* is "pitched", or inoculated, into the vat. Various species and strains of *Saccharomyces* are available for brewing. They are chosen for their unique genetic ability to ferment the wort at desirable rate and determine the character of beer. Brewery yeast are classified as either bottom-fermenting yeast (e.g., *S. carlsbergensis*) or top-fermenting yeast (e.g., *S. cerevisiae*). Bottom fermenters are used for the production of lager beer (the more common American beers) while top fermenters are used in the manufacture of ale. The fermentation usually lasts between seven to ten days (depending on whether it is top or bottom fermented) and is maintained at a temperature of about 40°C. As the fermentation proceeds, alcohol, acetic acid, glycerol, and carbon dioxide are formed. The carbon dioxide is piped from the fermenter, compressed, and stored. It is later returned to the beer during the carbonating process. When the fermentation is complete, the yeast population has increased six times and settled to the bottom of the tank. A portion of this yeast is washed and reused for other fermentations, while the excess is dried and added to poultry and hog feed. These yeast may also be used in the fermentation of corn or sugar beets to produce alcohol for gasohol. The beer is pumped from the fermenter for aging (lagering), filtering, carbonating and packaging. Beer in bottles and cans is pasteurized to lengthen its shelf life, but beer in kegs is not pasteurized and must be refrigerated.

In order to maintain production of a high quality beer, quality control measures are utilized throughout the brewing process. Equipment is cleaned and disinfected on a regular basis. Laboratory personnel maintain pure, active cultures of yeast, and a special effort is made to select for reuse only those yeast cells that will develop the highest quality beer. The final product is regularly checked to identify the presence of contaminating bacteria that may be responsible for beer defects.

## Distilled Alcohol as Beverage and Fuel

Alcohol produced during fermentation does not accumulate in the culture above a level of about 18 to 21 per cent, since it is toxic to the yeast producing it. For this reason, a simple fermentation does not result in a beverage with an alcohol concentration greater than 18 to 21 per cent. (The exact percentage depends on the strain of *Saccharomyces* used). In order to produce an alcoholic beverage with a higher concentration, *i.e.*, hard *liquor*, the beverage must be distilled from the original fermentation. *Distillation* is a heating process that drives alcohol vapour and some of its dissolved components from water and condenses it to a liquid that can then be purified, filtered, and clarified for consumption. Among the first alcoholic beverages distilled into liquors were wines and beer. Today a variety of liquors are made by distilling alcoholic beverages fermented from many types of starchbased grains, and sugary fruits and berries. Each beverage is made from special starting material, using fermentation techniques unique to each type of liquor. For example, Scotch whiskey is made from barley malt that has been kiln dried at a high temperature over a peat moss fire. This gives the whiskey its unique flavour. Bourbon is made from barley malt that has between 51 and 80 per cent added

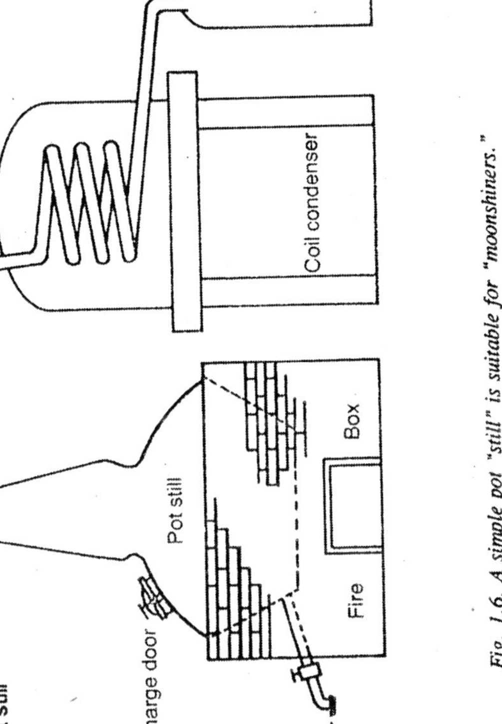


*Fig. 1.6. A simple pot "still" is suitable for "moonshiners."*

corn, and some rye is always added. Canadian whiskeys are fermented mainly from rye. The final concentration of alcohol in a liquor (given in "proof" number) is controlled by the distillation process and dilution with water. Packaged liquors such as vodka and gin are usually sold with a final proof of 80 or 90, but may be as high as 170 proof. The proof number is twice the alcohol concentration, *i.e.*, 170 proof vodka has an alcohol concentration of 85 per cent. Grain alcohol has a higher final proof, and its sale is regulated by state liquor departments. In most cases, this alcohol can only be purchased from a druggist and only with a valid letter of authorization describing its intended use.

**Table 1.1. Example of distilled alcoholic beverages.**

| *Name* | *Starch base* |
|---|---|
| Scotch | Barley |
| Bourbon | Corn, rye, barley |
| Rye | Rye and barley |
| Vodka | Potato |
| Irish whiskey | Rye, wheat, barley |
| Gin | Grains or potatoes with juniper berries |
| Arrak | Rice |
| Cognac | Grape |
| Brandy | Grape |
| Sloe Gin | Plum |
| Tequila | Agave (century plant) |
| Rum | Sugar cane |
| Applejack | Apples |
| Toddy | Coconut milk |
| Framboise | Raspberries |
| Liqueur and cordial | Mixture of distilled alcoholic beverages, herbs, sugar, and flavorings. |

In recent years the energy shortage has rekindled interest in ethyl alcohol as a liquid fuel to be used in automobiles, home heating systems and elsewhere. The idea is not really old because alcohol lamps, stoves, and many other appliances have been fueled with clean-burning alcohol for hundreds of years. Ethyl alcohol may be used as a liquid fuel in combination with gasoline as gasohol or as the more purified grain alcohol (190 proof). As an engine fuel, grain alcohol has several advantageous characteristics. It can be burned in an engine with only minor modification and with the efficiency of a high octane gasoline that prevent engine "knock". The combustion is so complete that noxious exhaust emissions are much lower and less toxic than those released from gasoline. Engines operating on alcohol run from 20° to 40°F cooler, extending their life and reducing the chances of overheating. Some research with alcohol/water fuel mixtures has demonstrated as much as a 16 percent increase in mileage. Currently, industrial fermentation plants designed for production of fuel alcohol are being constructed in many states. It is hoped that this will reduce the need for petroleum-derived gasoline by augmenting liquid fuel with alcohol, which is a renewable resource. An additional advantage to this type of fuel production is the generation of large amounts of high protein "spent" grain that can be used to supplement animal feed. Currently, research is underway to determine the average improved weight gain and health of farm animals fed this by-product of fermentation. There is strong indication that the increase will be as great as 10-20 percent.

There are four general types of distilled liquor: brandy, from fermented fruit juices; rum, from fermented molasses, whiskey, from fermented masses made with single types of the grains; neutral spirits, from fermented mash of mixed grains. In making whiskey and neutral spirits the grain, mixed with water, is autoclaved, cooled, diluted, and 1 per cent barley *malt* (aqueous extiact or sprouted barley) is added to hydrolyze the starch and proteins of the grain. The "mashing," or hydrolysis, proceeds in a special tank at about 65°C, for about 30 minutes. The mash is then pumped to the fermentation tanks. Here, as in beermaking, it is heavily inoculated with a starter of selected year, which has been cultivated in a mash previously made somewhat acid (pH 4.0) with lactobacilli. Fermentation is complete in about 72 hours, as in industrial alcohol production. The mash is then removed to the distillery and the ethanol, with various by-products, is recovered.

**Production of Butanol**

The production of butanol is outlined as an example of an industrial fermentation based on a species of bacterium.

As is true of a industrial ethanol production, much butanol is now derived as a by-product of petroleum "cracking." However, the biological process is still used to some extent and illustrates important microbiological principles. There are numerous species of *Clostidium* that ferment carbohydrates with the production of butyl alcohol and other materials of value in drugs, paints, synthetic rubber, explosives and plastics. Some species produce isopropyl alcohol and acetone as well. Important among these organisms are *Cl. acetobutylicum* and *Cl. felsineum*. The name of another species suggests its potentialities as an industrial agent: *Clostridium amylosaccharobutylpropylicum*!

Many wastes are rich in fermentable carbohydrates, *e.g.*. cannery refuse. Complete sterilization of all apparatus is essential. Conditions cannot be kept as acid (and antibacterial) as they are in yeast fermentations because *Clostridium* has its optimum pH near 7.2. Particularly troublesome contaminants are species of *Lactobacillus*. An organism called *B. volutans*, a grampositive, non spore forming rod (possibly a species of *Lactobacillus*?), is also especially dangerous.

Fermentation proceeds anaerobically for about three days. Normally butyl alcohol, acetone and ethyl alcohol, with carbon dioxide and hydrogen in large amounts, predominate when *Cl. acetobutylicum* acts in a mash rich in glucose. Other substances may occur in smaller amounts. Riboflavin (a vitamin of the B complex) in a valuable constituent of the residue after distillation of the fermental mash. Butyl and isopropyl (rubbing) alcohols are important among the volatile fermentation products of a related species, *Cl. butylicum*.

The successive reactions in the production of butanol and various side-products from glucose are as follows:

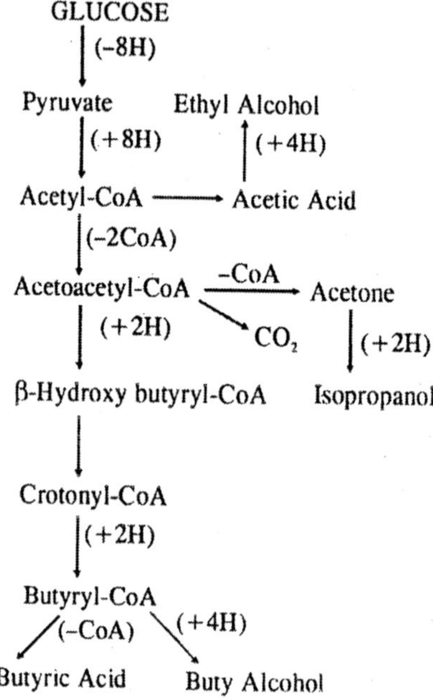


Lactic acid is commonly produced from lactose or glucose by species of homo-fermentative lactic acid bacteria, notably *Streptococcus lactis* or *Lactobacillus delbrueckii*;

$$\underset{\text{Lactose}}{C_{12}H_{12}O_{11}} + H_2O \longrightarrow \underset{\text{Glucose and Glucose}}{2C_6H_{12}O_6} \longrightarrow$$

$$\underset{\text{Pyruvate}}{4C_3H_6O_3} \xrightarrow{4H_2} \underset{\text{Lactic acid}}{4CH_3.CHOH.COOH}$$

Heterofermentative species of lactobacilli like *L. buchneri* purpose lactic acid and several by-products:

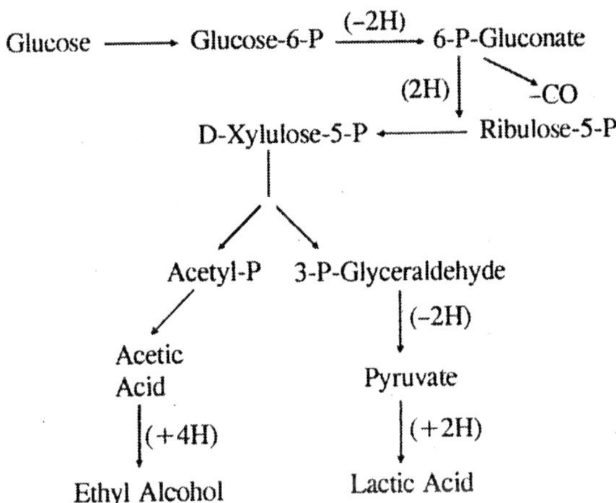


**Production of Vinegar**

Acetic acid is almost entirely responsible for the sour taste of vinegar. Indeed, a slightly sweetened, three per cent, aqueous solution of acetic acid makes a reasonable substitute for vinegar.

The occurrence of vinegar in fermented fruit juices was known to the ancients, although they had not knowledge of its cause. The bacteria involved were called *Mycoderma aceti* in 1862 by Pasteur.

The acid of natural vinegar is derived from alcohol by the oxidative action of bacteria of the Family Pseudomonadaceae (Genus *Acetobacter*). Pleasant flavours of natural vinegar are given by traces of various esters like ethyl acetate, and by alcohol, sugars, glycerin and volatile oils produced in small amounts by microbial action. Flavours are also derived from the fermented fruit juice, malt, or other alcohol liquor (wine, beer, hard cider) from with the vinegar was made.

In commercial vinegar-making by biological methods, preliminary fermentation of fruit juices to produce the necessary alcohol is often carried out by means of *Saccharomyces cerevisiae* (brewers' yeast). The *Acetobacter* then utilize the alcohol as a source of energy, oxidizing it to acetic acid in the presence of air. They utilize other substances in fermented liquor as foods. The alcoholic fluid is aerated as mucn as possible by various devices. In vinegar" generators" the alcoholic liquor trickles over the surface of aerated shavings, coke, gravel or other finely divided material inoculated with *Acetobacter*. Such an arrangement is called a *two-phase continuous process*; one phase is the down-trickling alcoholic liquor, the other phase is the column of coke or other material covered with growth of *Acetobacter*.

***Genus Acctobacter***

These are nonsporeforming, polar or peritrichous flagellate, gram-negative rods about 0.5 by 8.0 $\mu$, although species vary in size. Branching involution forms and large swollen cells frequently occur, especially in *mother-of-vinegar*, the gummy or slimy growth-phase of the organisms sometimes seen in natural vinegar or sour cider. Various species are found in souring fruits and vegetables. A species of historical interest is *Acetobactor (Mycoderma) aceti*, originally used by Pasture to demonstrate the biological nature of vinegar formation. In practice, several species of *Acetobacter* usually act jointly. The alcoholic and acidic nature of the process suppresses most contaminants. The overall reactions probably are as follow:

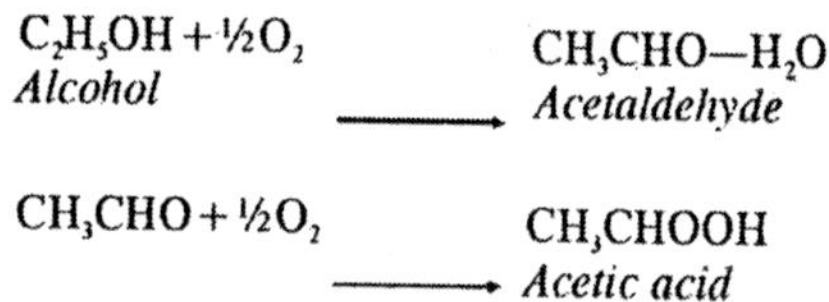

$$\underset{\textit{Alcohol}}{C_2H_5OH + \tfrac{1}{2}O_2} \longrightarrow \underset{\textit{Acetaldehyde}}{CH_3CHO—H_2O}$$

$$CH_3CHO + \tfrac{1}{2}O_2 \longrightarrow \underset{\textit{Acetic acid}}{CH_3CHOOH}$$

In a generator the rapid oxidation of alcohol by the organisms produces so much heat that careful control of the internal temperature by cooling coils is necessary.

## Foods From Waste

In the paper-pulp industry, wood chips are cooked for 6 to 18 hours at 60°C in solution of calcium bisulphite with free sulphur dioxide. The waste *sulphite liquor*, after the cooking process and removal of the wood fibers for papers, contains much valuable wood sugar (largely *xylose*) and other extractives. These form a good nutrient for asporogenous yeast or *Torula*.

The nutrients in such a medium may be turned into masses of yeast by adjustment of pH to about 5.0, removal of $SO_2$, aeration, additioin of nitrogen and phosphorus as $(NH_4)2HPO_4$ and $NH_4OH$, and inoculation with *Torulopsis utilis*. *Aerobic* growth is induced in aerated vats so that alcohol is not produced. The separation, drying and pressing of the resulting yeast growth are mechanical details. Yields of up to 50 per cent of the total reducing sugar consumed, in terms of dry torula, are obtainable. (At present, attempts are being made to cultivate yeasts on petroleum wastes). The yeast cells are rich in proteins, facts and vitamins. These are fed to stock or poultry and thus turned into meat and dairy products. Surely the transformation of a knotty old pine slab into a succulent pork chop or a fired egg is modern magic!

**Amino Acid Production**

Not only may yeasts serve as foods themselves, but some species, especially *Saccharomyces cerevisiae* and *Torula utilis*, while growing can synthesize large amounts of various amino acids. e.g., L-lysine. This is an expensive amino acid widely used to "fortify" many familiar foodstruffs. It is requisite that the growth medium for the yeast contain L-adipic acid or its derivatives. The yeasts use the adipic acid derivatives as *precursors* (*i.e.*, as the molecular raw material) for their synthesis of L-lysine.

***Hydrocarbons for protein***

Various hydrocarbon (petroleum) wastes are metabolizable by certain yeasts, eucaryotic fungi and also by some bacteria, *e.g.*, *Bacillus* spp., especially thermophilic species. One difficulty (or possibly a source of profit?) is the production of large amounts of heat by biooxidation of hydrocarbons. The proteins produced in the process are of high nuritive value and the method is potentially profitable.

## Microbial Enzymes in Industry

The enzyme industry is large and growing as more information relating to microbial metabolism becomes available through research. The industrial microbiologist applies this information in an attempt to increase the yield of a profitable product. Because these applications may be patented, much of the detailed information surrounding the actual growth of microbes and harvesting of enzymes is kept secret.

Enzymes are protein catalysts that speed the rate of specific chemical reactions and are produced through a complex sequence of protein synthesis reactions inside a living cell. The exact nature of an enzyme and its action is determined by the sequence of nucleotides found on the DNA of the cell in which it is manufactured. This fact makes it possible to select a particular genetic type of microbe for the production of an enzyme. The highly specific nature of enzyme action and the speed with which enzyme increase reactions make them excellent chemicals in many industrial, commercial, and medical processes. Microbes are excellent sources of enzyme, since they are easily grown in large aerated and submerged cultures, and produce enzymes on a regular basis. While in culture, bacteria, fungi, and filamentous bacteria synthesize and excrete *exoenzymes* that accumulate in the surrounding medium. Other enzymes are called *endoenzymes* since they are synthesized and utilized by the microbe inside the cell. To harvest these enzymes, the cultured cells must be separated, washed, and fragmented before the enzyme is isolated and purified.

Enzymes are highly specific in their actions and must chemically "fit" their substrate in order to function properly. Therefore, it is essential that they be produced and utilized in an environment that does not distort their shape or interfere with their action. One of the most widely used proteolytic enzymes that is frequently misused is found in household enzyme pre-soaks or as they are sometimes called "enzyme detergents." These enzymes are produced from cultures of the bacterium *Bacillus subtilis* and are able to hydrolyze many different proteins found in stains caused by chocolate, blood, and many foods. However, many people do not gain the maximum advantage from these products because they do not follow the product's directions, which are based on fundamental enzyme chemistry. These products should not be used in extremely hot water, because the high temperatures will denature the enzymes and prevent them from combining with substrate stains. Cold water slows their action and requires longer contact between the enzyme and substrate for effective stain-removal. By rubbing the enzyme directly into the stain, more effective and rapid contact is made and hydrolysis of the stain is more efficient. Since enzymes also lose their effectiveness in environments which contain inhibitors, they should not be used in certain types of "hard" water (water containing inhibiting minerals) or along with detergents such as the dodecyl-benesulphonate group, which includes most granular laundry detergents. Caution should also be taken with enzyme pre-soak products, since many people find them to be allergens. Wearing clothes cleaned in these products may cause allergic skin reactions Table 1.2 lists several other microbially produced enzymes used in many industrial, commercial, and medical processes and products. Their use must be carefully controlled. While both bacteria and fungi are used in the production of enzymes, it is the fungi that are primarily responsible for the production of large amounts of industrially important organic acids. They are used to produce resins which are, in turn, used for the manufacture of such items as plastics, varnishes, electrical non-conductors and medicines. Other organic acids are used for the production of such products as perfumes, oil additives, and textile dyes.

**Table 1.2. Some industrially produced microbial enzymes and their uses.**

| *Enzyme* | *Microbe cultured (Genus)* | *Uses* |
|---|---|---|
| Pectinase | *Aspergillus* | Separate fibres of the flax plant that may be used to make linen; same process used to make hemp rope. Used in green coffee processing. Used to clarify fruit juices. |
| Protease | *Aspergillus*<br>*Bacillus*<br>*Streptomyces* | Meat tenderizer; digestive aid; removes gelatin from photographic film for silver recovery; used in enzyme presoaks and detergents. |
| Amylase | *Bacillus* | Clarifies syrups; glucose production; softens dough in commercial bread making; removes starch and sizing from textiles; ingested as digestive aid, wallpaper remover. |
| Collagenase | *Bacillus*<br>*Clostridium* | Digests necrotic tissue from burns. |
| Lipase | *Rhizopus* | Digestive aid. |
| Rennin | *Mucor* | Curd formation in cheese-making. |
| Cellulase | *Trichoderma* | Digestive aid. |
| | *Streptomyces* | "Sweating" of animal hides to remove hair before tanning to leather. |
| Invertase | *Saccharinyies* | Soft-centered candies. |
| Urease | Fungal | Urea determination. |
| Lactase | *Lactobacillus* | Whole milk concentrates; lactose removal from milk for lactose-intolerant persons. |
| Streptokinase and streptodorase | *Streptococcus* | Digests necrotic tissue from burns. |

## Vitamins and Amino Acids

Any molecule that is required in minute amounts for the efficient operation of an organism and must be supplied from the environment since it is unable to be synthesized by that organism is known as a growth factor. Two of the most important classes of growth factors are the amino acids and vitamins. Although humans lack the ability to

**Table 1.3. Microbially produced organic acid.**

| *Acid* | *Microbe (Genus)* | *Use* |
|---|---|---|
| Lactic acid | *Lactobacillus* | Food preservative; in baking powder; as calcium lactate for food supplement for calcium deficiencies; bone development in pregnancies; used to make plastics. |
| Gluconic acid | *Aspergillus* *Penicillium* | Washing and softening agent; leavening agent; textile printing; As calcium glucontate for calcium deficiencies and in chicken feed to harden egg shells. |
| Citric acid | *Aspergillus* | In foods and beverages for taste and preservation. |
| Itaconic acid | *Aspergillus* | In manufacture of acrylic resins; plastic products; oil additives; dental resins. |
| Gibberellic acid | *Gibberella* *Fusarium* | Plant growth and flowering stimulant. |
| Kojic acid | *Aspergillus* | Analytical reagent; insecticide. |
| Fumaric acid | *Rhizopus* | Used as wetting agent and in manufacture of resins. |
| Gallic acid | *Aspergillus* | Ink and dye manufacture; skin disease cream. |
| Ustilagic acid | *Ustilago* | Used in "musks" for the perfume industry. |

produce some of these compounds, many microbes are able to synthesize them. Therefore, what is a growth factor to a human may be only a normal metabolic by-product to these microbes. Despite the fact that many microbes have the ability to synthesize these complex compounds, few produce excess amounts since they are essential to the normal metabolism of the microbe and are, in most cases, quickly utilized. Microbes selected for the industrial production of human vitamins and amino acids are defective mutants that have been isolated from wild cultures using special methods. These mutants lack the ability to use all the amino acids and vitamins they produce. The compounds accumulate inside the cells and are then released into the surrounding culture medium where they can be isolated and purified.

One of the most widely used amino acids produced industrially is glutamic acid. This amino acid is converted to the familiar compound monosodium glutamate (MSG) and added to many foods as a flavour enhancer. It is sold under the trade names Accent, Glutavene and Zest. MSG is added to many commercially prepared foods including baby foods, canned soups, meats, seafoods, and some soy sauces. MSG has been identified as a cause of brain damage; acute degenerative lesions have occurred in the retina of infant mice because of their age-dependent ability to absorb this compound. Although as yet unconfirmed, many researchers suspect that MSG may have a comparable ability to harm human infants. Since this food additive confers no significant benefit and, as yet, no confirmed harm to human infants, the Food and Drug Administration cautiously permits its unrestricted use, except for foods specifically designed for infants, However, many researchers question the safety of this chemical and continue to discourage its widespread use, especially in light of the fact that MSG has been confirmed to cause harm to adults. In 1968, it was discovered that this additive was responsible for Chinese restaurant syndrome, a disease characterized by headache, burning sensations facial pressure, and chest pain. These symptoms occur in adult after eating Chinese food prepared with high concentrations of MSG to intensify their flavours. In addition to glutamic acid, industrial microbiologists have produced other amino acids.

The greatest variety of human vitamins are manufactured by the yeasts *Saccharomyces cerevisiae* and *torula utilis*. These are sold in the form of compressed ironized yeast tablets and contain a mixture of many vitamins. The purified forms of vitamins are produced from cultures of bacteria such as *Streptomyces*, *Bacillus* and *Propionibacterium*. One of the most important is vitamins $B_{12}$ (cobalamin) which is produced in excess by *Streptomyces* growing in glucose or corn-steep broth with traces of cobalt. Vitamin $B_2$ (riboflavin) is produced from cultures of the fungus *Ashbya gossypii*, Although this fungus is a cotton plant pathogen, its vitamin product is one of the most widely used vitamin supplements in the commercial food industry.

## Antibiotics and Steroids

Antibiotics are unique molecules produced by microorganisms and they are able to destroy or inhibit the growth of other microbes. The industrial growth of large, submerged and aerated cultures of fungi, filamentous bacteria, and true bacteria marked a major break-through in the ability of the medical profession to treat and prevent infectious diseases. By culturing antibiotic producing microbes in these large fermentation vessels, each cell is able to come in contact with nutrients essential to the production of the drug and maximize its output. The search for and detection of antibiotic-producing microbes relies on samples taken from water, soil, and other materials from many parts of the world. In one method, microbes suspected of being antibiotic producers are isolated from mixed samples and plated on nutrient media for growth in pure culture. After lawns of growth are formed on the surfces of nutrient agar plates, small plugs are removed and placed on test plates inoculated with specific strains of *Staphylococcus aureus* or other pathogens. If any of the experimental microbes produce an antibiotic that is capable of destroying or inhibiting the pathogen, zones of clearing will form around the plug. These zones resemble those produced in a culture and susceptibility test performed in a medical laboratory. The microbes producing antibiotic are then growth in pure culture for further research to explore the nature of their metabolism and the likelihood that they will produce enough antibiotic to make them industrially profitable.

From agar slants the microbes are transferred to larger flasks to increase the number of cells. These are shaken continuously to ensure optimum accretion and growth. The shaker flask cultures serve as inocula for "bazookas" (transfer culture carriers) that are larger culture vessels used to increase the amount of inoculum to 10 per cent of the fermenter volume. Once inside the fermenter, the microbes grow, reproduce, and release significant amounts of the antibiotic. When the maximum amount of antibiotic has been produced, the solid components of the culture (*i.e.*, cells and other debris) are separated from the medium in a large rotary filter. The fermentation broth is then passed through a solvent extraction process that further separates the drug from unwanted chemicals. It is then transferred to an ionexchange chromatography apparatus that contains resins which adsorb the antibiotic to their surfaces. This process separates the drug from the remaining impurities before the antibiotic is crystallized under rigidly controlled conditions. The drug is then packaged in the form of capsules, tablets, liquids, or injectables. However, an alternative course can be taken to chemically modify the pure drug to other forms. This is done primarily to avoid the problem of drug resistance. As a result, entire families of antibiotics have produced from the fundamental compounds. The brief description of this process in no way reflects the years of intensive work needed to reach the point at which a microbe may be cultured for its valuable, government-approved products. In some cases, 10,000 or more species of strains of microbes have to be investigated before one suitable for culturing is identified, and years of research and field testing are necessary before the drug is made available to the public. It is also important to realize that of the over 2000 antibiotics described in research literature, only a few are of economic and medical value. The rest are far too toxic or ineffective to be used in the treatment of infectious diseases.

The last medically important group of compounds the steroids, are produced by an industrial fermentation process known as bioconversion. *Bioconversion* is the use of microbial fermentation to enzymatically change one compound to another biologically active from, in this complex series of fermentations, biologically inactive steroid compounds from plants are used as starting materials. A specific microbe is selected for its genetic ability to produce an enzyme that specifically alters the compound to another form. A series of several fermentations using different microbes is required to enzymatically change one compound into an other. The final product is a tailor-made molecule with desired biological activity. Some of the medically important compounds produced by bioconversion include bile salts, progesterone, testosterone, vitamin D, desoxycholate, cholesterol, estradiol, and cortisone. Among the more important genera of microbes used for bioconversions are *Rhizopus*, *Streptomyces*, *Aspergillus*, *Corynebacterium* and *Curvularia*. The complex fermentations required to convert one of the more basic compounds, progesterone, into a variety of other important molecules.

**Microbiological Assay**

Microbiological assay is a highly specialized application of the fact that certain organisms lack certain specific synthetic power, *i.e.*, are auxotrophs. *Lactobacillus plantarum*, for example, is unable to synthesize nicotinic acid ("niacin"). We may furnish the organism with a medium that is complete and satisfactory in all other respects but if niacin is lacking, absolutely no growth occurs. (Humans are no better off; without niacin they die of pellagra). If a minute amount (say, 0.01 microgram) per milliliter of niacin is added to the medium for *L. plantarum*, some growth will occur. More growth will occur in the presence of more of the missing factor. Up to the point of satiation or acidification, growth bears a linear relationship to the amount of the specific growth factor added.

For example, to assay the nicotinic acid content of fresh green beans, we prepare a medium for *L. plantarum* that is complete in all respects except niacin. This we omit. We now prepare two series (A and B) of 10 sterile tubes each. Each tube receives 10 ml. of the niacin-deficient medium. To each tube in series A we add known and graded amounts of pure niacin. To each tube in series B we add graded amounts of bean extract, niacin content unknown. All tubes are now inoculated with carefully washed (niacin free!) cells of *Lactobacillus plantarum*. Accurate, photometric measurements are then made of the growths (turbidities) obtained in the cultures. If the medium contains glucase, titrations of acidity instead of turbidity may be used as a measure of growth. By comparing growths in series A and B it is possible to estimate closely the concentration of niacin in the green beans. This method of estimation of growth factors is spoken of as *microbiological assay*.

Although the basic principle of all microbiological assays is the same, there are other methods of measuring the growth (or other physiological) response. These affect the cultural methods used. A commonly used procedure is the measurement of carbondioxide produced by fermentation of sugar in the test medium. Yeast is routinely used in the microbiological assay of *thiamine* (vitamin $B_1$) by this method. In *pyridoxine* (vitamin $B_6$) assays, the mold *Neurospora* is the test organism. after sufficient incubation the culture is steamed and the entire mycelium of *Neurospora* is removed from the culture medium dried and weighed. Dry weight is directly proportional to concentration of pyridoxine in the sample of material being assayed. Another assay procedure depends on the spherophast-producing power of the assayed substance.

Certain organisms lend themselves very well to such assay procedures. *Lactobacillus casei* and *L. arabinosus* are easy to cultivate, relatively hardy, harmless and wholly dependent upon several growth-factors including various amino acids, riboflavin, biotin, pantothenic acid and nicotinic acid. Other organisms may be used for assay for other substances, for example, *Streptococcus lactis* of folic acid. Ultraviolet-induced, synthetically deficient auxotrophs of molds, yeasts ad bacteria are extremely valuable in assay work.

Even though the basic principle of microbiological assay is easily understood, the technological details are often exceedingly complex and filled with pitfalls. Many obscure factors affect the test organisms, and they may also undergo mutation and other changes without notice. Mutational and other injuries may be held to a minimum by storage of the stock cultures in containers with liquid nitrogen at –196°C. (–321°F). Temperature, pH and presence or absence of air may be of critical importance. For example, under *aerobic* conditions, *Lactobacillus lactis* will *die* before it will grow without vitamin $B_{12}$! *Anaerobically*, it sneers at vitamin $B_{12}$! There are many other examples. We may smile, but knowledge of this and many other peculiarities is essential to successful assay procedures.

## Industrial Spoilage

In contrast with the useful activities of bacteria, a word may be said of their destructive action. Several causes of industrial spoilage (*e.g.*, "diseases" of fermentations) have been mentioned in this chapter and in the chapters on soil, food and water bacteria. Species of *Micrococcus, Alkaligenes flavobacterium, Serratia, Clostridium*, coliform organisms, yeasts and molds are common causes of spoilage.

Each type of product is attacked by certain species of micro-organisms that can metabolize the substance especially well. For example, spoilage of cellulosic products such as lumber, telephone poles; paper; sisal, jute and flax fibers; tobacco and cotton is brought about by cellulose decomposers such as molds, various species of *Clostridium*, *Cellulomonas*, *Cytophaga* and many other such organisms of the soil. Fermentable substances such as syrups and beverages are attacked by yeasts, lactobacilli, organisms of the coli-aerogenes group and various environmental bacteria including the Genus *Clostridium*. Spoilage of proteins such as meats, fish, milk and so on result from the action of proteolytic species such as *Pseudomonas*, *Bacillus*, *Proteus*, *Micrococcus*, *Clostridium* and many others. Petroleum hydrocarbons are attacked by certain soil bacteria, as already mentioned, and rubber insulation of vital communication wires is attacked by bacteria and eucaryotic fungi.

Lactobacilli and *leuconostoc* species have already been noted as particular villains in the acid-food, fermentation and distillery industries. Species of both can ruin fruit or vegetable juice or various industrial mashes (beer and wine) during processing. They produce a buttermilk flavour. Pasteur found *Lactobacillus* and *Leuconostoc* causing "diseases" of beers and wines. They are just as active today. Lactobacilli also discolour meats, especially producing greenish discolouration (oxidized porphyrins) or cured hams and sausages.

The slimy dextran or levan-forming species, such as *Leuconostoc mesenteroides* and *L. Dextranicum* and some lactobacilli and micrococci, produce slimy and ropy conditions in a great variety of human endeavours; sugar refineries, pickle brines, dairy products, ham-curing cellar, and the like. These organisms prefer acidified products such as partly fermented foods, mashes and citrus juice. Examples of several of these types of spoilage have been given in discussions of the various products.

Development of undesirable flavours in fatty products such as butter, especially rancidity, is due in great part to the formation of butyric acid as a result of lipolysis It is caused by species of *Aspergillus* and other molds, *Pseudomonas* species and streptococci related to *Str. liquefaciens*. These difficulties do not arise when clean equipment, clean milk and proper precautions to avoid contamination are used.

Proteolytic organisms, such as *Str. liquesfaciens* are responsible for undesirable bitter flavours and early spoilage of cheeses and other protein products. Gas production is usually caused by coliforms and *Clostridium*; putrefaction or digestion by *Clostridium*, *Pseudomonas* and *Bacillus*. Such conditions result mainly from dirty milk or other food equipment, or careless handling.

Included among other sabotage activities of bacteria are corrosion of the inside of structural aluminum alloys used in fuel tanks of jet-fuel aircraft. Pitting and scaling of the metal occurs beneath heavy, slimy growths of hydrocarbon-utilizing bacteria such as species of *Pseudomonas* and *Desulphovibrio*. Some species of molds are also involved. Bacteria of the Genera *Mycobaterium* and *Nocardia*, among others, are also implicated in the deterioration of bituminous products, including asphalt highways and asphalt coatings and pipe-linings. The micro–organisms seem to utilize the high viscosity hydrocarbons and resins in asphalt.

## Prevention of Spoilage

This, in each instance, is a problem that can be solved only by careful examinations of the process involved to find: (a) the nature of the organism(s) involved; (b) where the contamination is getting in, and (c) then devising means of excluding it. It is impossible to lay down a blanket rule for industrial spoilage in general. Everything depends on maintaining conditions unfavourable to, or excluding by asepsis, organisms that can grow on or in the particular product involved. This may involve complete steam sterilization of fermentation equipment (tanks, pipes, pumps); drying; refrigeration; aeration; the use of inhibitory salt, sugar or acid concentrations' radiation with ultraviolet light; exposure to sunlight; treatment with substances such as creosote, sodium benzoate and the like. In some processes specific antibiotics may be used, as penicillin and tetracyclines in alcoholic fermentation of molasses.

# 2

# Biogas Generation and Comparisons

Various methods of methane generation from biomass are described including: complete mix and stratified regimes, anaerobic contact process, upflow anaerobic sludge blants, rotating biological contractors, packed columns and supported growth fluidized bed. The advantages and limitations of each are discussed and specific applications are suggested. In general, suspended growth systems are preferable for wastes with high solids content and supported growth systems are indicated for soluble organics and, in some case, smaller volumes. The character, volume and location of the waste will usually determine the choice of system. Pretreatment methods for enhancing gas production and digestion temperatures are also discussed.

Natural gas, which is mostly methane, is the cleanest, most nonpolluting fuel currently in use. It provides about 25% of the energy in the United States today, heating about one-half of the homes and running clothes dryers, air conditioners, dishwashers and a whole range of appliances for which the only substitute is electricity, that can be five or even ten times as expensive. The methane delivery system consists of 350,000 miles of underground high-pressure pipelines, 650,000 miles of distribution mains, and connections to 45 million users. This system, which is also connected to a storage system of exhausted underground wells, is capable of carrying a volume of gas considerably larger than present usage. Methane from biomass can be fed into this system, as it is being done from feed lots in Oklahoma and a landfill in California.

A choice of the most economical method for the destruction of an organic waste material depends on many factors, including the capital investment in equipment required as well as operating costs and energy requirements for aeration, pumping, heating, sludge dewatering and credits for usable or saleable byproducts.

In aerobic treatment, the organics are generally oxidized to carbon dioxide via the citric acid cycle, producing a high energy yield. Usually this results in rapid cell growth and substrate degradation, with a concurrent large production of sludge. In anaerobic digestion, much of the chemical energy in the substrate is retained in the methane produced. Only limited amounts of energy are available for cell growth, resulting in low growth rates and a small sludge production. Due to the slower growth rates of the methane forming bacteria the hydraulic residence time as well as the mean cell residence time is greater than that for aerobic digestion. With cell recycle, anaerobic treatment is possible even with highly dilute wastes; however, the economics of building larger reactors must be considered.

There seems to be general agreement that wastes with BOD (*biochemical oxygen demand*) > 3000 and COD (*chemical oxygen demand*) > 4000 are more economically treated by anaerobic than aerobic methods, and at higher concentrations (> 20,000 mg/I COD) anaerobic methods cost about 25% of equivalent aerobic methods. An upper economic limit of COD strength for anaerobic digestion has not been established, but it is probable that at concentrations > 50,000 mg/I evaporation and solids recovery may well compete with anaerobic digestion.

In summary, advantages of the anaerobic process as compared with the aerobic process are:

1. Energy recovery is possible through the use of methane generated.
2. Higher organic loadings are possible.
3. A smaller volume of sludge is produced.
4. Energy consuming aeration is unnecessary.
5. It can be cost effective at loadings between 4000 and 50,000 mg/l COD.

Disadvantages are:

1. Digesters (or influent) must be heated.
2. The process is somewhat more susceptible to upset.
3. The slower growth rate of methane bacteria requires longer detention times.

4. The process may not be economical for dilute wastes (BOD < 3000 mg/l).

In individual cases, whether or not anaerobic treatment and energy recovery is feasible depends on many secondary factors including:

1. Quantity of waste to be treated,
2. Continued availability of waste (e.g., fruit and vegetable cannery wastes are often only seasonably available),
3. Toxic materials in waste stream that may require specific pretreatment procedures,
4. Pretreatment that may be required to enhance biodegradability,
5. Physical characteristics of waste that may pose special problems such as foaming, flotation, etc., and
6. Transportation costs to the treatment facility.

## Description of Systems

Anaerobic digestion systems may be divided into two general categories: suspended growth, that is where the microorganisms are suspended in the liquid of the reactor, and supported growth systems where the biomass grows as a film attached to some sort of support media. The character of the waste will probably suggest the most appropriate system(s) design.

## Suspended Growth Systems

Suspended growth systems may be operated in unmixed (stratified) or mixed regimes, or a combination of both in series. When solids are recycled from a secondary stratified reactor to a primary complete mix reactor the system is known as the "contact" process. Upflow anaerobic sludge blanket reactors are also suspended growth systems.

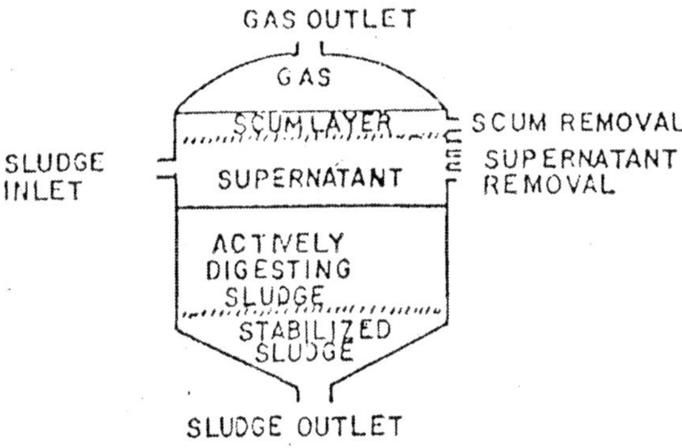


*Fig. 2.1. Standard rate (stratified) digester.*

Unstirred (conventional rate) digesters are stratified, and may be heated or unheated. Solids loading rates are usually 0.03-0.10 lbVSS/$ft^3$/day and they are operated as batch reactors with intermittent feeding and withdrawal. Detention times are 30-60 days. Variations of this system have been designed for rural, agricultural installations.

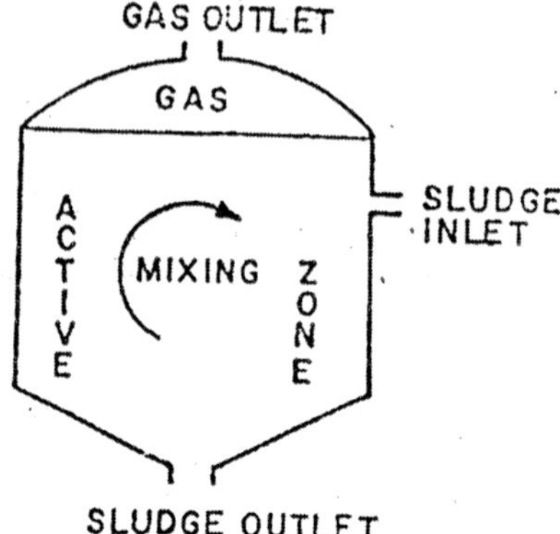


*Fig. 2.2. High rate (complete mix) digester.*

Complete mix (high rate) digesters may be operated at higher loading rates, 0.1-0.2 lb VSS/$ft^3$ /day, and shorter detention times, 15-20 days. They are usually heated (30-35°C) and operated in a continuous mode. First order rate constants based on carbon concentrations have been found to be 0.086/day as compared to 0.054/day in unstirred reactors. Multiple stirred reactors (up to 40) placed in series have demonstrated high methane yields at short detention times. Complete mix and unstirred reactors may also be operated in series, with the stratified reactor effecting solids separation. When a portion of the

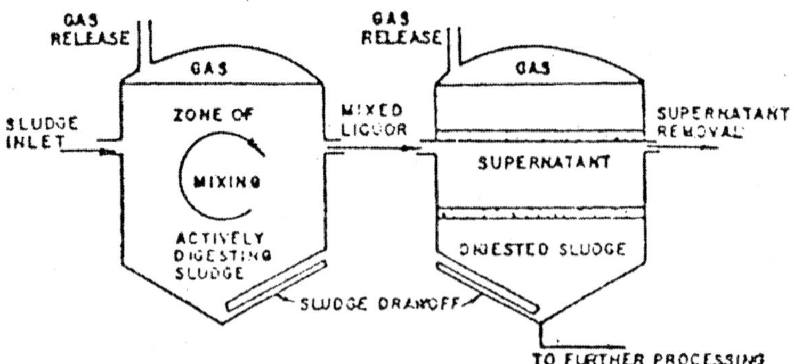


*Fig. 2.3. Two-stage anaerobic digester.*

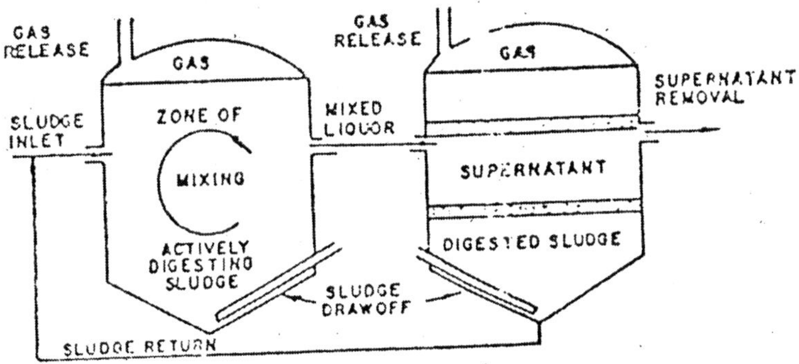


*Fig. 2.4. Anaerobic contact digester.*

settled sludge is recirculated to the first tanks (as in the activated sludge process) the system is called an anaerobic contact process. This practice increases the rate of waste stabilization and gas production. A critical solids retention time of 10 days has been suggested.

The *upflow anaerobic sludge blanket* (USAB) was developed based on the concept that anaerobic sludge inherently has superior settling characteristics. This is probably a result of the shape of the particles (oval), and their relatively high density probably caused by the presence of ferrous sulfide. This is important to prevent "wash-out" of the reactor. Mixing is effected by the circulation of flow and by rising gas bubbles. Settlers are mounted inside the reactor. Flow in the

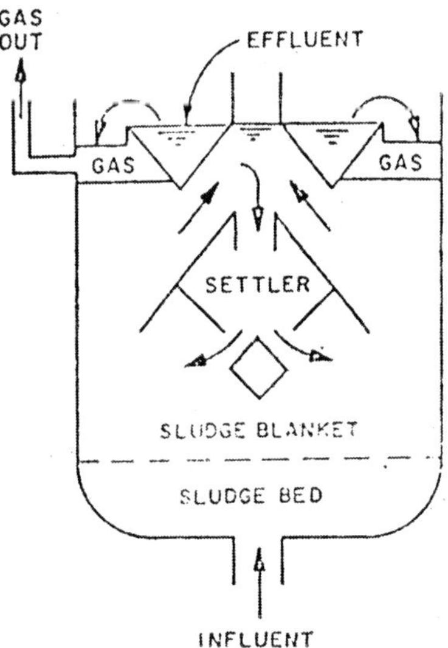


*Fig. 2.5. Upflow sludge blanket reactor.*

settlers is laminar and retention time is low to prevent gas formation; the thickened sludge is constantly being fed back into the reactor. USAB reactors are able to tolerate high organic and hydraulic loadings without process failure. Overloading takes place when the linear fluid velocity in the settler exceeds the settling velocity of the sludge, or if the recirculation of the sludge is exceeded by the sludge input into the settler. Removal efficiencies of 90% at loading rates of 10 to 25 kg/ $m^3$-day and hydraulic retention times of 5-6 hrs. at 30°C have been suggested.

**Supported Growth Systems**

Supported growth systems are those in which the micro-organisms grow as a biomass attached in some manner within the reactor. This provides for longer solids (microbial) retention times, a well acclimated biomass and a resistance to wash-out at low hydraulic retention times. The attached film of biomass gradually increases in thickness as the organics are metabolized until nutrients can no longer diffuse to the organisms in contact with the support. That portion of the biofilm will then be sloughed off and carried into the effluent. For this reason a final settling chamber is usually required. This chamber need not be large, for the settling characteristics of this sludge are excellent. The thickness to which the film may grow before sloughing is, to some extent, controlled by the hydraulic regime (i.e., shear stresses) in the reactor. Types of attached growth systems include submerged rotating biological contactors, upflow anaerobic filters and fluidized beds.

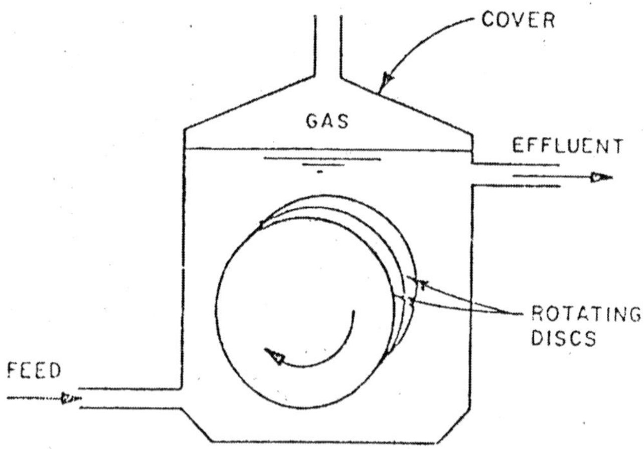


*Fig. 2.6. Submerged rotating contractor.*

Rotating biological contactors consist of many rotating corrugated polyethylene discs. The discs are usually relatively small (50 cm) to minimize the velocity differential along the face; however, several units may be operated in series. A typical rotational velocities is 13 rpm for a 47 cm diameter disc. Active biomass is attached to the reactor walls as well as to the discs, and substrate removal is proportional to biomass area. These units have been shown to be effective in the treatment of high-strength soluble substrates and hydraulic retention times of 4.4 to 8.8 hours have been suggested.

Upflow anaerobic filters provide a support media for the biofilm. Gravel, Raschid rings, limestone chips, rough glass or polystyrene beads, and even oyster shells have been used. Oyster shell and limestone packing contribute buffering capacity tc the system. Good removals have been reported for high-strength (COD 10,000 mg/l) wastes. Substrate removal increases with increased *hydraulic retention times* (HRT). HRT's of 10 to 24 hrs. usually effected COD removals between 70 and 80% while HRT's of 10 days or more 97% COD removals have been reported. Recirculation is often used to increase HRT and protect the system from shock loads as well. Addition of surfactants has been reported to increase gas production by as much as 40%, possibly by enhancing the separation of gas bubbles from the biomass and/or by activating the cell membrane. Up flow filters have been suggested for secondary treatment of municipal waste waters as well as for high strength commercial soluble waste streams. These systems may be operated between 20°-26°C as well as in the mesophillic range.

The *fluidized bed biofilm reactor* (FBBR) is a supported growth system in which the micro-organisms grow as a film on a free moving carrier particle. In such a reactor the available surface area per unit of reactor volume is greater than in either packed bed or rotating disc systems. Wastewater is passed upward through a bed of media such as sand, charcoal or synthetic material at velocities sufficient to fluidize the media. The only variables that can be controlled by the design engineer are media characteristics such as size and density, as well as column height and allowable expansion. These parameters will, in turn, determine up flow velocity, porosity and biofilm thickness. For a given set of operating conditions there is an optimum media size and density. Too-small media cause excessive bed expansion and a decrease in biomass concentration and thus in the rate of substrate removal, while too-large media size decreases the available surface area and biomass concentration. Increasing the density of the media enhances substrate removal, since for a given upflow velocity, less bed expansion and therefore greater biomass concentration will result. This advantage must be weighed against the higher energy requirements for fluidizing heavier media.

These systems have been shown to be capable of achieving high organic removal efficiencies at low temperatures (10-20°C) and treating relatively lowstrength (COD-600 mg/I) at short HRT's (several hours) and high organic loading rates (up to 8 kg. COD/$m^3$-d). High substrate concentrations can sometimes result in bed flotation, suggesting that diffusion of gaseous products may be rate limiting, for if gas production rates exceed diffusion, gas bubbles will be formed, disrupting the system. Recirculation may be used to "dilute" high COD wastes. As with other supported growth systems FBBR's require that the waste streams be low in suspended solids.

## Applications

### Treatment Plant Sludges

Wastewater sludge is being generated in enormous quantities at sewage treatment plants, particularly at activated sludge facilities. Recent regulations have mandated both the end of ocean dumping of sludges and provisions for full secondary treatment. These regulations will result in increased production of sludge and necessity to treat and dispose of it in an acceptable manner. Anaerobic digestion is one of the processes employed in the stabilization of these sludges, to remove from the raw sludge its odour, pathogens, putrescibility and other offensive characteristics.

Using the methane produced by the anaerobic digestion of sludge to supply power for the sewage treatment plant is not new. This practice was used in the 1940's and the 1950's by many municipalities, but was gradually abandoned in the 1960's when electricity became inexpensive. Aerobic digesters were chosen in place of anaerobic ones because of their relative ease of operation and resistance to upset. However, they were energy users. Increased energy costs have resulted in a renewed interest in anaerobic sludge digestion since the methane produced can often both heat the digester and supply the bulk of the power needed by the entire treatment plant.

Due to the high solids content of the sludges suspended growth systems are necessary, and sludge digesters currently in use in the United States are either stratified, complete mix, contact stabilization, alone or in combination. Although supernatant and thermal pretreatment liquors are generally returned to the head of the treatment plant, upflow filters have been suggested to ease the load on stressed facilities.

Since sludge handling may represent as much as 30-40% of the capital cost and 50% of the operating cost of a treatment plant, investigations to optimize the process continue. Thermal pretreatment (180°C for 30 min.) results in improved biodegradability (more gas, less sludge), improved dewaterability, odour control and sterilization. Thermal pretreatment prior to anaerobic digestion may actually result in an increase in net energy production, based on expected increase in biodegradability and hence in gas production.

The addition of activated carbon to the digesters (45 kg/day) improved the overall treatment efficiency and allowed less chlorine to be used. Savings in filter chemicals, labor and chlorine more than paid for the carbon used. Processes have also been investigated using a two-stage combination of anaerobic and aerobic steps using solar energy for heating requirements and under pressure of two to three atmospheres.

The principal use of digester gas is heating the digester. It has been estimated that approximately 30% of the gas production is used for direct heating in temperature climates. The amount of gas necessary to heat a digester depends on many factors, including insulation, siting, climate and exposure. A conventional hot water heat exchanger system using direct gas heating is about 70% efficient, although use of waste heat from a gas engine, turbine or other source would make the digester gas available for other uses. Two of the main London sewage works, Beckton and Mogden, were producing 2.22 and 2.03 million $ft^3$ gas/day in 1971-1972.

A small amount of this gas is purified and sold, but nearly all is used to run dual-fuel engines or gas turbines driving air compressors and generators which provide 80-90% of the total power needed in the sewage works.

An $83 million Yonkers Joint Treatment plant in Westchester County, New York, is demonstrating the cost effectiveness of using waste heat and sludge gas. Sludge gas is used to produce steam and heat and to fuel dual-fuel engines (92% methane, 8% diesel oil) which drive process blowers. Of the gas produced, 90% is used, of which 10% goes to produce steam and 90% to fuel engines. Other conservation efforts at this plant are related to heat recovery. Engine exhaust gas (1200°F) is used to generate steam, contributing 20% of the total building heating requirements. Energy recovered from engine jacket and lubricating oil (via heat exchangers) is used to heat digesters. Fuel savings at Yonkers are estimated to be about $600,000/yr.

**Feed Lot Wastes**

Modern intensive farming systems where animals are kept in feed lots are ideal candidates for methane from animal waste production systems. The amounts of wastes produced by farm animals varies, but typical figures are much higher per capita than for humans, so that a farm of 1000 pigs has a sewage disposal problem equivalent to a town of 4000 people. The amounts of gas produced per pound of dry-weight waste will vary with the type and age of the animal as well as with the diets. Animal wastes generally contain more lignocellulose material than domestic wastes and thus more resistant to attach by micro-organisms. The percent of organics digested anaerobically has been reported as 50-70% for poultry wastes, 50-60% for swine and 10-26% for dairy cows. The digestability of pig and cattle manure may be increased by 170% by sodium hydroxide pretreatment. This treatment consists of adding approximately 8% of a 50% NaOH solution and allowing it to reactor for 14 days before neutralization and digestion. Digestion proceeds satisfactorily, although more slowly at temperatures below the mesophillic range. The economics of larger reactors vs. heating costs in a particular climate will determine the best design.

A system for the conversion of animal wastes to methane called "Anox" is valued at £30,000 and can digest about 120 $m^3$ of pig manure with a digester volume of 1500 $m^3$. The gas is used to generate electricity at the rate of 1 kW/hr/150 pigs. The solids are dewatered and used for fertilizer and the supernatant is treated with ozone. Economic evaluation indicates an annual cost of £62,250 as compared to credits of £96,099.

The Calorific project in Guyman, Oklahoma, processes about 500 ton/day of cattle manure producing not only liquid fertilizer and methane (160,000 $ft^3$/day which is upgraded to pipeline quality) but also cattle feed from the dried solids. Studies conducted to evaluate the efficacy or recovered biomass as a high protein feed have found it to be comparable to soybean meal as a feed supplement.

Experiments to determine optimum operating conditions suggest the following:

1. Digesters cannot be started up using "neat" feces and urine. Microbial seed from other digesters must first be acclimated using slurries of approximately 5% volatile solids. As the microorganisms become acclimated, solids loading can be increased.

2. Detention times of stabilized digesters should be maintained at about 10 days. If the SRT falls below 7 days, performance can drop off rapidly.
3. 35°C is a reasonable digestion temperature (mesophilic). Digesters adapted to 30°C exhibited performances about 5% below those at 35°C. Sudden changes in temperature are upsetting to performance.
4. Scum formation can be a major problem. Scum in farm waste digesters consists of animal hairs, etc., and fibrous feed residues which, when not mixed in, dry on top of the digester contents to form an impermeable layer. In one case a scum of hen feathers blocked the top of the digester and pipes and caused an explosion. Scum formation can be prevented by stirring digester contents.
5. Scroll and flexible stator pumps have been found best for farm waste which is more abrasive on pumps and pipes than is domestic sewage sludge.
6. The methane produced is sufficient to heat digesters, stir digesters and operate pumps.
7. In-ground holding and digester tanks are more thermally efficient.
8. Other wastes, when available, can be added to the digester to increase gas production. These include silage effluent, rotten potatoes, crop residues and waste vegetable matter.
9. Of the number of possible designs for farm digester plants, the single-stage, stirred tanks are the most useful. Two stage or feed-back systems would not appear to justify the increased cost and energy requirements.
10. Thermal pretreatment of the animal wastes might be indicated if an economic use for the additional methane produced is available.
11. Shapes of digestion tanks greatly influenced circulation of tank contents and scum formations. Egg-or-pear-shaped European digesters have fewer scum problems and better circulation than cylindrical digesters, however, they are more expensive to construct.

**Rural Installations**

Small package plants have been designed and are in use in small villages in undeveloped countries. Large number of these plants are in use in China and more than 20,000 have been installed in India over the past 15 years. One design used in Taiwan, South Korea, India and elsewhere has two digester compartments. The first is about 3' 8" × 3' 8" × 6' and the second 8' 2" × 4' 10" × 6'. This unit was

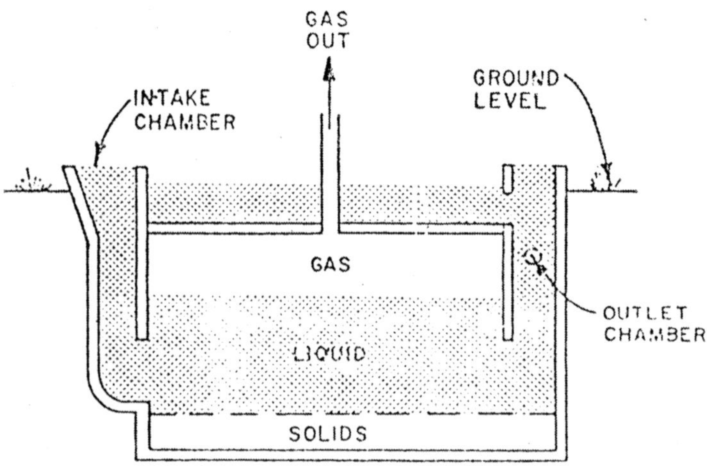


*Fig: 2.7. Typical rural biogas reactor.*

designed for a 20 pig farm and is said to provide fuel for cooking for a family of twelve "if operated properly." The efficiency of these units varies considerably, not only with the skill or the operators, but also with climatic conditions.

This type of reactor can be built and operated with unskilled labor. Generally, the input and proportions are human wastes 10%, animal feces 30%, and crop stalks 10%, mixed in 50% water. It has been found that if the moisture content drops below 45% gas production decreases substantially. Before plant matter is put into the digester, it should be composted for at least 10 days so that decomposition will have begun.

These reactors are installed in-ground and are not heated so that gas production rates vary seasonally. During summer and autumn digester temperatures average 23°C and considerable gas is produced. In winter, with ambient temperatures ranging from 0°C to 7°C, digester temperatures stay at about 10 °C and gas is produced more slowly. The unit should be desludged once a year. If the removed sludge is to be used as fertilizer it should be stabilized at pH 12 to destroy pathogens and parasitic eggs.

**Food Processing Wastes**

Disposal of processing wastes is a major problem for the food processing industry. Nearly all types of food processing wastes have been shown to be suitable substrates for methane generation. Suspended growth systems are more appropriate for wastes with high solids while supported growth systems may be preferable for soluble or settled wastes. A major drawback, in many cases, is the seasonal nature of the wastes. Duration of the "season" for any particular crop is, typically, one or two months. Methane producing reactors are not amenable to changes in substrate and usually require an acclimation period of at least a few weeks before gas production resumes. For this reason, construction of elaborate facilities may not be fiscally attractive; however, studies continue to determine how best to change from one digester feed to another without process disruption.

Promising results have been obtained from anaerobic digestion studies of brewery by-product. One of the advantages is the generation of a saleable microbial biomass, for the bacteria generated are nearly 70% protein. Additionally, since carbohydrates and lipids in the substrate decompose more quickly than do the proteins, the product is a more highly concentrated protein feed than was the substrate. The brewery industry in the United States has 185 facilities producing malt beverages including beer, ale and malt liquor. Wastewater is generated at the rate of 8.4 liters (excluding cooling water) per liter of beer produced. Average wastewater loadings for a large facility are 1,400 mg BOD/l and 570 mg SS/l. It is further proposed that since the cooling water has a temperature of approximately 70°C, it be used to heat the digesters.

Food processing and crop wastes are often deficient in nitrogen and phosphorus, and addition of these nutrients is required. Studies indicate that for maximum methane production a COD/N/P ratio of 300/5/1 is usually adequate.

## PRETREATMENT

Anaerobic digestion of primary and secondary sludges generally results in reductions of organic matter of about 50 and 35% respectively. If this efficiency could be improved, benefits would be accrued not only from the greater volume of methane produced, but also from the smaller quantity of sludge requiring disposal. Thermal pretreatment has been studied by several investigators and has been shown to have the following advantages:

1. Improved degradability and thus increased gas production,
2. Improved dewaterability,
3. Destruction of pathogens, and
4. Reduced sludge volumes remaining for disposal after digestion.

Temperatures between 150°C and 250 °C have been suggested, but a typical pretreatment program is 175°C for 30 minutes. At higher temperatures there is danger of the formation of toxic substances such as furan compounds, which inhibit microbial action. Addition of sodium hydroxide prior to thermal pretreatment often further enhances the degradability of the wastes. The effect of the pretreatment is the solubilization of as much as 60 to 70% of the suspended solids and the breakdown of structural lignocellulose materials that may physically restrict the extracellular hydrolizing enzytnes excreted by the microorganisms.

# 3

# ETHANOL PRODUCTION

Over the last 25 years there has been considerable interest in the potential of producing fuel ethanol from biomass. The OPEC oil embargo of the 1970s resulted in a marked increase in the price of oil and influenced countries such as Brazil, Canada, Finland, Japan, Sweden and the USA to plan for better liquid fuel self-sufficiency. More recently research into fuels from renewable resources has been driven by environmental concerns, particularly the role of fossil fuel contribution to poor air quality and global warming. As a result there has been considerable research and discussion about the environmental advantages of using ethanol as a transportation fuel and as a gasoline supplement.

The benefits of using ethanol and the ether form of ethanol, ethyl tert-butyl ether (ETBE), as an alternative to gasoline, have been discussed in detail. Ethanol is a clean-burning, high-octane fuel that can be readily substituted for gasoline and its combustion results in significant reductions of toxic emissions such as formaldehyde, benzene and 1,3-butadiene.

Blends of ethanol or ETBE with gasoline increase the octane of the mixture and can improve performance. Ethanol blends cause internal-combustion gasoline engines to run with leaner fuel mixtures and they reduce carbon monoxide emissions and it further lowers the Reid vapour pressure of gasoline, thereby decreasing the release of smog-forming compounds including ozone. Ozone is recognized as being one of the most pervasive and persistant urban air-quality problems. Consequently, urban areas in the USA with heavy air pollution, such as areas of California and Calorado, have led the way towards implementation of ethanol (and other oxygenated fuel) vehicle regulations.

When ethanol is produced from renewable sources such as biomass it can both decrease urban air pollution and reduce the accumulation of carbon dioxide, one of the greenhouse gases. Thus replacement of gasoline with ethanol, derived from renewable biomass feedstocks that sequester $CO_2$ during growth, is expected to reduce $CO_2$ emissions by 90-100 per cent. It has also been estimated that enough neat ethanol could be made from the cellulosic biomass residues that are currently available within the USA today to potentially replace twice the amount of gasoline consumed within the USA in 1994.

## Feedstocks

### Sugar and Starch

Ethanol, when used as a transportation fuel, can be generated from a number of feedstocks which are usually categorized into sugar, starch and lignocellulosic based materials. Sugar crops including sugar cane, sugar beets and sweet sorghum produce monomeric sugars (glucose, fructose and sucrose) that can be directly fermented to ethanol. Starch crops, including grains (corn, wheat, barley, grain sorghum) and tubers (potatoes, sweet potatoes) require an extra processing step, called hydrolysis, prior to fermentation. Hydrolysis converts the complex sugars or starches in the grains and tubers to monomeric sugars suitable for fermentation. Fuel ethanol is currently produced from both sugar and starch based feedstocks with approximately 3 billion US gallons of ethanol produced from sugar cane in Brazil, 1 billion US gallons from corn in the USA, and 5.5 million US gallons from both corn and wheat in Canada. However, the use of these feedstocks for fuel production competes with other higher value usages such as food production. Current production of fuel ethanol is often based on excess agricultural production and it is generally recognized that this volume is too small in comparison with the anticipated levels of production required for total conversion of transportation fuel markets from gasoline to ethanol. It is also apparent that there is the potential for competition with food production for both the sugar and starch feedstocks and that prime agricultural lands normally required to produce the foodstuffs should not be diverted for fuel production.

### Lignocellulosics

Lignocellulosic biomass is typically composed of a complex mixture of three polymers—cellulose, hemicellulose and lignin—and a small amount of other compounds that are loosely termed extractives. The fermentation of sugars derived from lignocellulosic feedstocks has proven to be more of a process design and operating challenge than traditional sugar or starch based processes. For example, there is a considerably wider variation in the type and nature of the processes and equipment needed to convert lignocellulosic feedstocks to ethanol. Although sugar, starch and lignocellulosic substrates can have compositional variability due to variations in the species of feedstock used, growing site, climate, age and the part of the plant used, lignocellulosic feedstocks have the following additional problems: proportional variability within the mixture of the three major components; differences in the types and amounts of extractives; and natural variability in the monomeric sugars that make up the hemicellulose component.

## Acid vs. Enzymatic Hydrolysis of Lignocellulosics

### Acid Hydrolysis of Lignocellulosics

Acid hydrolysis of biomass feedstocks has been studied and practised commercially for many years. Although several types of acid, including sulphurous, sulphuric, hydrochloric, hydrofluoric, phosphoric, nitric and formic have been used for hydrolysis, there are essentially two types of acid hydrolysis process, termed dilute and concentrated, with a number of genetic processes associated with each of these two options. Although the conversion of lignocellulosic to glucose via acid hydrolysis has been proven technically on a large commercial scale it is still considered that enzymatic hydrolysis has the potential to surpass greatly the efficiency of acid hydrolysis. Major technical problems that have yet to be resolved using acid hydrolysis are: the corrosion of the reaction vessels; degradation of product sugars resulting in low yields; the need for neutralization before subsequent bioconversion; formation of numerous environmentally noxious by-products; high capital and operating cost; and solvent losses. Although a considerable amount of research is still directed towards these problem areas, the long-term outlook for acid based biomass to ethanol processes is still not overly optimistic as various economic projections indicate that several major technical problems have yet to be resolved before the economics of an acid based process are significantly more attractive.

### Enzymatic Hydrolysis of Lignocellulosics

Enzymatic hydrolysis processes are relatively new and integrated processes have yet to be proven both technically and economically. Enzyme based hydrolysis of lignocellulosics tends to show better promise than acid based hydrolysis, primarily because of the potential for higher sugar yields and the production of less toxic effluent streams. However, any proposed enzyme based bioconversion process is generally more complex than an acid hydrolysis process as the enzymes tend to show more substrate specificity and require more carefully controlled reaction conditions. To date, a truly 'genetic' enzymatically-based biomass-to-ethanol process has been difficult to identify because of the heavy influence that the type of feedstock, type of by-products, number of unproven processes and equipment currently available will have on the design of such a process. However, it is generally acknowledged that a generic enzymatic-based process would include the following steps: pretreatment, fractionation, enzyme production, enzyme hydrolysis, fermentation, ethanol and other by-product recovery, and waste treatment. The pioneering work that has been done at the pilot or demonstration scale has shown that the subprocess steps are all strongly interdependent. Thus it has been extremely difficult to identify the relative technical or economic merits of each of the subprocess variations and their subsequent influence on the final production cost of ethanol.

Some of the process steps, such as fermentation and ethanol recovery, have been commercialized and often used as component steps in other industries such as brewery and distillery plants. These component process steps therefore have an established technoeconomic baseline and can be considered to be mature technologies. The less mature process steps—pretreatment, fractionation, hydrolysis and pentose fermentation—have generally been compared on a relative technical or economic basis using laboratory, pilot-plant equipment or techno-economic modelling to simulate the entire, integrated process. There have been a number of pilot plants, two fully-integrated pilot plants and several smaller non-integrated pilot plants built and operated over the last 5 years. They have been primarily used to test the technical and economic feasibility of various aspects of the enzymatic conversion process. There have also been a number of attempts to model techno-economically the various process scenarios using laboratory/pilot-plant equipment.

## Major Component Steps in an Enzyme Based Biomass-to-Ethanol Process

### Pretreatment

Pretreatment is the process step required to make the relatively recalcitrant lignocellulosic material more easily digestible to the hydrolytic enzymes while preserving the yield of the original carbohydrates for fermentation. This can be accomplished by various mechanisms such as the removal of the lignin sheath, reduction of cellulose crystallinity, or by increasing the surface area that is accessible to the enzymes. However, inhibitory breakdown products can be formed if the pretreatment conditions used are too harsh. The nature of the lignocellulosic substrate used has a major impact on this process step as a certain pretreatment, which may be effective with one lignocellulosic substrate, may prove to be ineffective on another. A number of reviews have covered pretreatment in detail and these have generally separated the different types of pretreatment into physical, chemical, biological and combinations of these methods. There appear to be four main pretreatment methods currently being researched and commercialized: organosolv, steam explosion, dilute-acid prehydrolysis. Of these various options, only the steam-explosion process has resulted in the substantial commercialization and sale of reactors by companies such as Stake Technology of Canada.

**Fractionation**

Fractionation is the subprocess step that separates the lignocellulosic slurry obtained after the pretreatment into the three main fractions of cellulose, hemicellulose and lignin. Subsequent fractionation after pretreatment is generally recognized as an efficient way of providing for separate processing of the individual components while recovering most of the material available in the original feedstock. There are generally two unit operations within the process which provide separation of both the hemicellulose and lignin components. The major product derived after these two fractionation steps is a cellulose rich residue which is subsequently hydrolyzed enzymatically. Although pretreatment and fractionation can be carried out simultaneously in processes such as organosolv, in processes such as steam explosion or acid hydrolysis it is usually carried out as two separate subprocess steps. Due to the inter-related nature of the enzymatic based biomass-to-ethanol process, the type of feedstock and pretreatment method used will probably determine the fractionation procedure that is adopted. However, the greater the number of washing steps required, the higher the capital and operating costs will become, and the more likely that the net return on the investment will be negative.

Hemicellulose, once it has been solubilized through pretreatment by $SO_2^-$ steaming or dilute acid, can be processed in a number of different ways. Currently, the xylose derived from agricultural or hardwood hemicelluloses cannot be readily fermented to ethanol. This will be discussed more fully in the subsequent fermentation section. Although there has been a considerable amount of work carried out on other hemicellulose derived products such as furfural, xylitol and single-cell protein this has not lead to any significant commercial products, partly because the few high-value products that were identified have too small a market volume.

After pretreatment, lignin can generally be extracted from most lignocellulosic residues by a sodium hydroxide wash. Lignin has a high heat content and has been used traditionally in the pulp and paper industry as a boiler fuel for process heat or cogeneration of steam and electricity. The potential for by-product utilization is immense because approximately 1 kg of lignin is produced per litre of ethanol. Although there have been a large number of high-value lignin-derived products identified, the operation of a few commercial-scale plants would probably saturate the world market for most of these applications.

**Hydrolysis**

The enzymatic hydrolysis of the cellulosic component of lignocellulosics requires the use of a complete cellulase enzyme complex containing various endoglucanases, exoglucanases and cellobiases. Various groups, due primarily to technical and economical reasons, have advocated the on-site production of cellulases rather than using the commercial cellulases sold by companies such as Novo-Nordisk, Genencor, Primalco and at least a further seven companies worldwide. These companies currently produce and market different types of cellulases for applications in areas such as textiles, detergents, animal feed and pulp and paper processing. Cellulases are synthesized and excreted by various microorganisms with certain fungi, primarily the *Trichoderma* species (notably *T. reesei*), among the most efficient producers. Generally, the amount of cellulase produced by wild-type strains of fungi or bacteria is too low to support an economical industrial process. For this reason strain improvement programmes were initiated in the mid-to-late 1970s and still continue to this day. Strains originally isolated using agar plating and isolation techniques have been improved by various methods, such as increasing the production of the whole cellulase complexes, increasing the resistance to glucose repression and enhancing pH and temperature tolerance. Certain strains, such as *T. reesei* have remained remarkably stable, in a genetic sense, over the whole transition from laboratory-scale to precommercial production scale (30,000-litre fermenter). However, continued genetic improvement through both molecular biology and further use of traditional mutagenesis and screening protocols has not significantly improved the multifactorial characters such as specific activity, productivity or yields.

Cellulase production is now possible at a cost well below expectations based on earlier economic studies. However, the cost of supplying enzymes for a biomass-to-ethanol process is still too high to allow the economic production of chemicals or fuels from lignocellulosic derived sugars. Although on-site production of hemicellulose and glucose hydrolysates could provide a cheap source of substrates, the cost and effectiveness of on-site enzyme production has yet to be proven. Consequently, a significant amount of work on the production, mechanism and effectiveness of cellulase is continuing in this area. It can therefore be expected that, with increasing volumes of enzyme scales and the increased number of new applications in areas such as pulp and paper, the commercial cost of enzymes will continue to drop.

As mentioned earlier, enzymatic hydrolysis is accomplished through the synergistic action of several cellulase components. Synergistic action means that the combined activity of the enzymes is greater than the sum of each of the components. The current mechanistic model primarily involves three main groups of enzymes that are required before cellulose can be hydrolyzed effectively to glucose. The first component includes the endoglucanases, which attack the amorphous cellulose in a random action, producing more free which remove cellobiose units from both the reducing and non-reducing ends of cellulose chains. The third component includes the β-glucosidases which split the cellobiose units into monomeric glucose units. Each of the enzyme components is influenced by end-product inhibition and consequently the build-up of any of the products from any of the enzymatic reactions results in inhibition of the overall cellulose hydrolysis reaction.

The cellulase enzyme complex has been isolated from a wide variety of organisms including anaerobic protozoa, aerobic fungi, and aerobic and anaerobic bacteria. Enzymes from *trichoderma reesei*, which is an aerobic, mesophilic fungus, are the most extensively studied cellulases, essentially because all the necessary enzyme components for cellulose hydrolysis are produced extracellularly in high concentrations and the organism can be grown in submerged aerobic culture. Furthermore, the cellulase complex produced is resistant to chemical inhibitors and remains stable for up to 48 hours at 50°C. Although maximum cellulase activity for most fungal derived cellulases occurs at 50 ± 50°C and pH of 4.0–5.0, cellulase complexes are known to differ substantially in their pH and temperature tolerance and in the ratio and amount of the different cellulase components. For example, *T. reesei* wild-type preparations are deficient in cellobiose, a b-glucosidase type enzyme, and result in an accumulation of cellobiose unless supplemented with the enzyme. However, there are induced mutant strains of *T. reesei* that display high β-glucosidase activity.

The high specificity of the cellulase enzymatic reaction should theoretically result in efficient hydrolysis of cellulose to glucose. In practice, the yields of glucose are influenced by many factors that can impact on both the rate and extent of hydrolysis. For example, it is known that the hydrolysis rate progressively declines with time due to a range of substrate and enzyme related factors. It has been shown that the surface area available for enzyme-substrate interaction is influenced by cellulose pore size and the shielding effects of hemicellulose and lignin. The crystalline structure of cellulose excludes water molecules as well as any larger molecules, including the cellulase enzymes, and thus reduces the available surface area. Although it has been suggested that the crystalline regions of cellulose will be hydrolyzed at a much slower rate than the amorphous cellulose, due to the greater stability resulting from the interchain hydrogen bonding, this has not proven to be the case. Analysis of residual substrates has shown that the crystallinity remains unchanged as hydrolysis proceeds..

As well as the surface area of the substrate limiting hydrolysis, both the rate and extent of the reaction are influenced by enzyme factors such as end-product inhibition, the irreversible and/or non-specific adsorption of cellulases onto the substrate, and the inactivation of key components of the cellulase complex. There are various ways in which the effectiveness of the cellulases could be enhanced. For example, the low specific activity of commercial cellulase preparations has led to the requirement for high cellulase enzyme loadings. As a result, a considerable amount of research is now focused on learning more about the enzymatic and inhibitory mechanisms associated with the cellulase complex, screening for strains with enhanced enzymatic features (higher productivity, yield, resistance to end-product inhibition and higher specific activity) and the designing and testing of alternative operational and flow characteristics (e.g., fed-batch reactors, high enzyme concentrations, enzyme recycling) to enhance the activity and reuse of the cellulases.

## Fermentation

### Hexose Fermentation

The fermentation of glucose to ethanol was one of the first complex biological and chemical processes mastered by man. Alcohols became an important fuel and chemical feedstock in the mid-nineteenth century, predating the petrochemical industries of today. For example, most solvents and chemicals such as acetone, butanol and ethanol were originally produced from a number of carbohydrate feedstocks using a variety of conversion processes. However, with the rapid growth of the petroleum and petrochemical industry following World War I, fermentation research and development has been restricted primarily to the brewing and distilling industries and the last two decades of fermentation research have tried to extend the limits of traditional technologies rather than develop radically different processes. By optimizing the typical batch cycle, a batch fermentation which took 7 days in 1978 was reduced to 3 days by 1986. At the same time companies such as Melle-Boinot developed a system which operated in a mode between a batch and continuous fermentation and reduced the fermentation time to 16 hours. This was accomplished by recycling the yeast and residual fermentation substrate from earlier batches, thereby decreasing the volume of the fermenter and consequently increasing yields. Simultaneous ethanol distillation and advances in antibiotic research have also aided in the development of continuous fermentation processes. The ability to draw off the ethanol continuously as it is produced enables the system to keep a high level of productivity by removing the product which can slow down the metabolism of the yeast and eventually kill it. Advances in antibiotics were also required as continuous fermentation systems are vulnerable to outbreaks of infection and subsequent costly shutdowns. The Biostil process incorporates many of these developments while other methods and technologies continue to be researched to increase the efficiency of ethanol production. Strategies include the development of flocculating yeasts, extractive fermentation, yeast immobilization, and continued research into modifying the various organisms through various classical and genetic engineering techniques.

**Pentose Fermentation**

The importance of being able to ferment the hemicellulose pentose derived sugars to the economics of the biomass-to-ethanol process has been emphasized by a number of authors. There are several reasons why the fermentation of pentoses to ethanol continues to be researched aggressively. These include the relative ease by which pentose can be recovered from most lignocellulosic substrates by methods such as dilute acid hydrolysis and acid catalyzed steam explosion; the rapidly expanding knowledge base on the biochemistry of ethanol production from xylose continues to offer potential ways of enhancing fermentation efficiency. Finally, as mentioned above, a major driving force is the economic incentive that the extra ethanol from the xylose fraction of many biomass sources has on the overall ethanol process. The identification of a xylose-fermenting yeast, *Pachysolen tannophilus*, in 1981 first indicated that ethanol could be derived from xylose and allowed projections to be made of the economic viability of converting both cellulose and hemicellulose derived sugars to ethanol. Other pentose fermenting organisms such as *Pichia stipitis* and *Candidae shihatae* have also been identified and characterized over the last 5 years. The focus of most of this research has been to understand the limitations of the xylose fermentation and to use modern genetic engineering techniques to improve the ethanol yields and productivity of the yeasts. Some recent work has looked at the application of xylose-fermenting organisms in processing schemes using actual lignocellulosic sugar substrates. This has indicated that various problems such as inhibitory products and fermentation of mixed sugar streams have yet to be fully resolved.

Several groups have been trying to manipulate certain bacteria and yeasts genetically so that they could incorporate the genes necessary for xylose fermentation to ethanol. Metabolic engineering has been applied to *Erwina chrysanthemi*, *Escherichia coli*, *Klebsiella oxytoca*, *Klebsiella planticola* and *Zymomonas mobilis* and to *Saccharomyces cerevisiae*. So far the organisms which are most capable of fermenting both hexoses and pentoses have been *E. coli*, *K. oxytoca*, *Z. mobilis* and *S. cerevisiae*. Enteric bacteria such as *E. coli* do not grow well on xylose, even though they possess the necessary genes for xylose uptake and utilization. Normally, anaerobic fermentation of sugars with *E. coli* results in a range of products including lactate, acetate, succinate and formate with ethanol being only a minor product. Although engineered organisms can directly ferment xylose and other five-carbon sugars they tend to prefer a neutral pH. However, a neutral pH creates a greater demand for base and a higher potential for contamination by other organisms.

**Ethanol Recovery**

The beer resulting from a typical glucose fermentation usually contains about 8-12 per cent ethanol by volume. This dilute concentration is primarily a consequence of ethanol end-product inhibition. It is not possible to produce anhydrous ethanol by simple distillation as an azeotrope is formed at 95% ethanol; the azeotrope has a lower vapour pressure than either ethanol or water and is therefore preferentially distilled. A second step using a third component (e.g. benzene) to form another azeotrope with one or both of the original components has traditionally been employed. A considerable amount of work continues to be carried out to try to reduce the effects of end-product inhibition and the costs of conventional distillation (vapour recompression, cascade pressurization, super-critical fluid carbon dioxide) and dehydration (carbon dioxide extraction, solvent extraction, extractive fermentation, pervaporation, molecular sieves, adsorption).

## Current and Future Status of Enzymatic Hydrolysis

As is apparent in this brief review, the bioconversion of lignocellulosics to ethanol is a complicated and strongly interdependent series of process steps. Currently there are no true examples of commercial or totally integrated demonstration-sized plants which can convert lignocellulosic materials to ethanol. Although some of the current and past pilot plants have been able to demonstrate the successful operation of entire sets of subprocess steps, these plants have not been able to operate continuously over a prolonged period of time. Hopefully, plants such as those at NREL in Colorado, USA, and others that will be potentially constructed in Sweden and Canada will provide the technical and operational experience that will allow the establishment of true commercial plants.

As a less expensive option, many researchers have attempted to assess the current techno-economic status and future potential of the various bioconversion processes by building techno-economic models based on information obtained from both laboratory and pilot studies and from experience gained in the operation of similar processes in other industries, for example, ethanol from gain/corn. At this point in time, the comparison of these models appears to be the most direct way of providing a relative subprocess cost estimation. Although it is often difficult to compare model results directly as the basic tenets on which the models are based can differ substantially, there is general agreement that an enzyme based biomass-to-ethanol process could produce ethanol for about US$0.50 per litre based on the recovery of most of the cellulose and hemicellulose derived sugars and an energy credit for burned lignin. More optimistic studies incorporating advances in genetic engineering, ethanol recovery and by-product credits have estimated that US$0.13-0.18 per litre of ethanol can be achieved. Although continued research will undoubtedly decrease the cost of producing biomass-to-ethanol it is probable that the eventual increase, both environmentally and economically, in the cost of petroleum derived fuels will provide the necessary incentive to establish biomass-to-ethanol processes.

# 4

# FOOD PRESERVATION

The widest use of sorbates is as fungistats in products such as cheeses, bakery products, fruit juices, beverages, salad dressings, and the like. With molds, inhibition has been reported to be due to inhibition of the dehydrogenase enzyme system, and to the inhibition of cellular uptake of substrate molecules such as amino acids, phosphate, organic acids, and so on. A number of other possible inhibitory mechanisms have been presented by various researches. Against germinating endospores, sorbate prevents the outgrough of vegetative cells. With respect to toxicity, sorbic acid is metabolized in the body to $CO_2$ and $H_2O$ in the same manner as fatty acids normally found in foods.

## BENZOIC ACID AND ITS SALTS

Sodium benzoate was the first chemical preservative permitted in foods. Benzoic acid ($C_6H_5COOH$) and its sodium salt ($C_7H_5NaO_2$) along with the esters of P-hydroxybenzoic acid (parabens) are considered together in this section. Food and Drug Administration, and it continues in wide use today in a large number of foods. Its approved derivatives have structural formulas as noted:

Methylparaben
Methyl *p*-Hydroxybenzoate

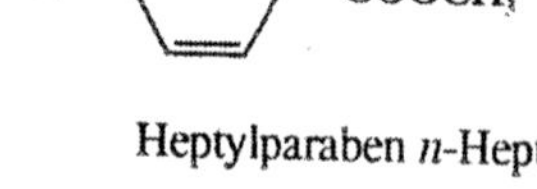

HO—⟨benzene ring⟩—COOCH$_3$

Propylparabt
Propyl *p*-Hydroxybenzoate

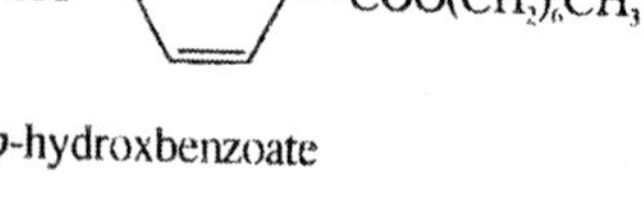

HO—⟨benzene ring⟩—COO(CH$_2$)$_6$CH$_3$

Heptylparaben *n*-Heptyl-*p*-hydroxbenzoate

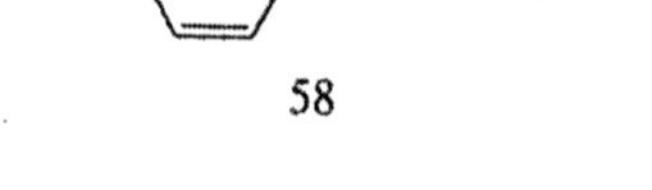

HO—⟨benzene ring⟩—COO(CH$_2$)$_6$CH$_3$

The antimicrobial activity of benzoate is related to pH, the greatest activity being at low pH values. The antimicrobial activity resides in the undissociated molecule. These compounds are most active at the lowest pH values of foods and essentially ineffective at neutral values. The pK of benzoate is 4.20 and at a pH of 4.000, 60% of the compound is undissociated, while at a pH of 6.0 only 1.5% is undissociated. This results in the restriction of benzoic acid· and its sodium salts to highacid products such as apple cider, soft drinks, tomato catsup, and salad dressings. High acidity alone is generally sufficient to prevent growth of bacteria is these foods, but not that of certain molds and yeasts. As used in acidic foods, benzoate acts essentially as a mold and yeast inhibitor although it is effective against some bacteria in the 50–500 ppm range. Against yeasts and molds at around pH 5.0–6.0, from 100–500 ppm are effective in inhibiting the former, while for the latter from 30–300 ppm are inhibitory.

In foods such as fruit juices, benzoates may impart disagreeable tastes at the maximum level of 0.1% The taste has been described as being "*peppery*" or *burning*. As noted above, the three parabens that are permissible in foods in the United States are heptyl-, methyl-, and propyl-, while butyl-, and ethylparabens are permitted in food in certain other countries. As esters of p-hydroxybenzoic acid, they differ from benzoate in their antimicrobial activity in being less sensitive to pH. Although not as many data have been presented on heptylparaben, it appears to be quite effective against microorganisms, with 10–100 ppm effecting complete inhibition of some gram-positive and gram-negative bacteria Propylparaben is more effective than methylparaben on a ppm basis, with up to 1,00 ppm of the former and 1,000–4,000 ppm of the

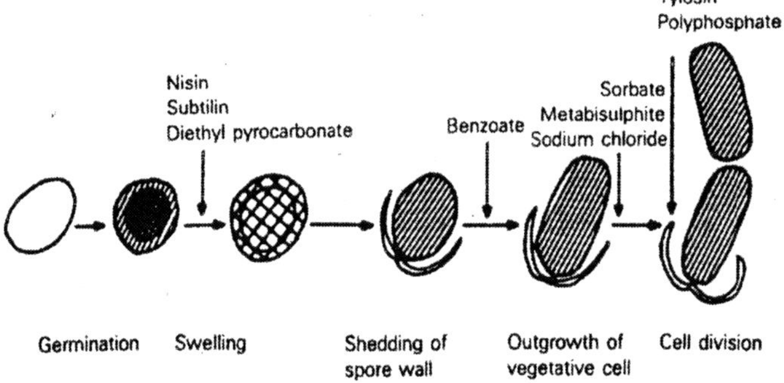


*Fig. 4.1. Diagrammatic representation of growth of an endospore into vegetative cells showing stages arrested by minimum inhibitory concentration of some food preservation.*

**Table 4.1. Summary of Some chemical food preservatives.**

| *Preservatives* | *Maximum tolerance* | *Organisms affected* | *Foods* |
|---|---|---|---|
| Propionic acid/propionates | 0.32% | Molds | Bread, cakes, some cheeses, rope inhibitor in bread dough. |
| Sorbic acid/sorbates | 0.2% | Molds | Hard cheeses, fogs, syrups, salad dressings, jellies, cakes. |
| Benzoic acid/benzoates | 0.1% | Yeasts and molds | Margarine, pickle relishes, apple cider, soft drinks, tomato catsup, salad dressings. |
| Parabens | 0.1% | Yeasts and molds | Bakery products, soft drinks, pickles, salad dressings. |
| $SO_2$/sulfites | 200-300 ppm | Insects, microorganisms | Molasses, dried fruits, wine making, lemon juice (not to be used in meats or other foods recognized as sources of thiamine). |
| Ethylene/propylene oxides | 700 ppm | Yeasts, molds, vermin | Fumigant for spices, nuts. |
| Sodium diacetate | 0.32% | Molds | Bread |
| Dehydroacetic acid | 65 ppm | Insects | Pesticide on strawberries, squash. |
| Sodium nitrite | 120 ppm | Clostridia | Meat curing preparations. |
| Caprylic acid | — | Molds | Cheese wraps. |
| Ethyl formate | 15-2 ppm | Yeasts and molds | Dried fruits, nuts. |

latter needed for bacterial inhibition, with gram-positive bacteria being more susceptible than gram negative to the parabens in general. Heptylparaben has been reported to be effective against the malolactic bacteria.

In a reducedbroth medium, 100 ppm propylparaben delayed germination and toxin production by *C. botulinuin* type A, while 200 ppm effected inhibition up to 120 h at 37°C. In the case of methyl-paraben, 1,200 ppm were required for inhibition similar to that for the propyl-analog. The parabens appear to be more effective against molds than against yeasts. As in case of bacteria, the propylderivative appears to be the most effective where 100 ppm or less are capable of inhibiting some yeasts and molds, while for heptyl and methylparabens, 50–200, and 200–1,000 ppm, respectively are required. Like benzoic acid and its sodium salt, the methyl- and propylparabens are permissible in foods salt, to 0.1% while heptylparaben is permitted in beers to a maximum of 12 pp, And up to 20 ppm in fruit drinks and beverages. The pK for these compounds is around 8.47, and their antimicrobial activity is not increased to the same degree as for benzoate with the lowering of pH as noted above. They have been reported to be effective at pH values up to 8.0. Form a more thorough review of these preservatives. Similarities between the modes of action of benzoic and salicvlic acids have been noted. Both compounds, when taken up by respiring microbial cells, were found to block the oxidation of glucose and pyruvate at the acetate level in *Proteus vulgaris*. With *P. vulgaris*, benzoic acid caused an increase in the rate of $O_2$ consumption during the first part of glucose oxidation.

The benzoates, like propionate and sorbate, have been shown to act against microorganisms by inhibiting the cellular uptake of substrate molecule. The undissociated form is essential to the antimicrobial activity of benzoate as well as for other lipophilic acids such as sorbate and propionate, as previously noted. In this state, these compounds are soluble in the cell membrane and act apparently as proton idnophores. As such, they facilitate proton leakage into cells and thereby increase energy output of cells to maintain their usual internal pH. With the disruption in membrane activity, amino acid transport is adversely affected.

### NaCl and Sugars

These compounds are grouped together because of the similarity in their modes of action in preserving foods. NaCl has been employed as a food preservation since ancient times. The early food uses of salt were for the purpose of preserving meats. This use is based upon the fact that at high concentrations, salt exerts a drying effect upon both food and microorganisms.

Nonmarine microorganisms may be thought of as normally possessing a degree of intracellular tonicity equivalent to that produced by about 0.85–0.9% NaCl. When microbial cells are suspended in salt (saline) of this concentration, the suspending menstrum can be said to he isotonic with respect to the cells. Since amounts of NaCl and water are equal on both sides of the cell membrane, water moves across the cell membranes equally in both directions. When microbial cells are suspended in, say, a 5% saline solution, the concentration of water is greater inside the cells than outside (concentration of $H_2O$ is highest where solute concentration is lowest). It should be recalled that in diffusion, water moves from its area of high concentration to its area of low concentration. In this case, water would pass out of the cells at a greater rate than it would enter. The result to the cell is *plasmolysis*, which results in growth inhibition and possibly death. This is essentially. what is achieved when high concentrations of salt are added to fresh meats for the purpose of preservation. Both the microbial cells and those of the meat undergo plasmolysis (shrinkage), resulting in the drying of the meat as well as inhibition or death of microbial cells. To be effective, one must use enough salt to effect *hypertonic* conditions. The higher the concentration, the greater the preservative and drying effects. In the absence of refrigeration, fish and other meat may be effectively preserved by salting. The inhibitory effects of salt are not dependent upon pH as are some other chemical preservatives.

Most nonmarine bacteria can be inhibited by 20% or less of NaCl, while some molds generally tolerate higher levels. Organisms that can grow in the presence of and require high concentrations of salt are referred to as *halophiles*, while those that can withstand but not grow in high concentrations are referred to as *halodurics*. The interaction of salt with nitrite and other agents in the inhibition of *C. botulinum* is discussed above under nitrites. Sugars, such as sucrose, exert their preserving effect in essentially the same manner as salt. One of the main differences is in relative concentrations. It generally requires about six times more sucrose than NaCl to effect the same degree of inhibition.

The most common uses of sugars as preserving agents are in the making of fruit preserves, candies, condensed milk, and the like. The shelf-stability of certain pies, cakes, and other such products is due in large part to the preserving effect of high concentrations of sugar, which like salt makes water unavailable to microorganisms. Microorganisms differ in their response to hypertonic concentrations of sugars, with yeasts and molds can grow in the presence of as much as 60% sucrose while most bacteria are inhibited by much lower levels. Organisms that are able to grow in high concentrations of sugars are designated *osmophiles*, while *osmoduric* microorganisms are those that are unable to grow but are able to withstand high levels of sugars. Some osmophilic yeasts such as *Saccharomyces rouxii* can grow in the presence of extremely high concentrations of sugars.

**Sulfur Dioxide and Sulfites**

Sulfur dioxide ($SO_2$) and the sodium and potassium salts of sulfite ($=SO_3$), ($-HSO_3$), and metabisulfite ($=S_2O_5$), and all appear to act similarly and are here treated together. Sulfur dioxide is used in its gaseous or liquid forms, or in the form of one of more of its neutral or acid salts on dried fruits, in lemon juice, molasses, wines, fruits juices, and others. The parent compound has been used as a food preservative since ancient times. Its use as a meat preservative in the United States dates back to at least 1813; however, it is not permitted in meats or other foods recognizable as sources of thiamine. While $SO_2$, possesses antimicrobial activity, it is also used in certain foods as an antioxidant. The predominant ionic species sulfurous acid depends upon pH of milieu with $SO_4$ being favoured by pH$<$3.0, $HSO_3^-$ by pi between 3.0 and 5.0, and SO3 > pH 6.0. $SO_2$ has of 1.76 and 72. The sulfites react with various food constituents including nucleotides, sugars, disulfide bonds, and others. With regard to its effect on microorganisms, $SO_2$ is bacteriostatic against *Acetobacter* spp. and the lactic acid bacteria at low pH, concentrations of 100 to 200 ppm being effective in fruit juices and beverages. It is bactericidal at higher concentrations. When added to temperature-abused comminuted pork, 100 ppm of $SO_2$ or higher were require to effect significant inhibition of spores of *C. botulinum* at target levels of 100 spores/ g.

The source of $SO_2$ was sodium metabisulfite. Employing the same salt to achieve an $SO_2$ concentration of 600 ppm, Banks and Board found that growth of salmonellae and other Enterobacteriaceae were inhibited in British fresh sausage. The most sensitive bacteria were eight salmonellae serovars, which were inhibited by 15–109 ppm at pH 7.0, while *Serratia liquefaciens*, *S. marcescens*, and *Hafinia alvei* were the most resistant, requiring 185–270 ppm free $SO_2$ in broth.

Yeasts are intermediate to acetic and lactic acid bacteria and molds in their sensitivity to $SO_2$, and the more strongly aerobic species are generally more sensitive than the more fermentative species. Sulfurous acid at levels of 0.2–20 ppm was effective against some yeasts, including *Saccharomyces*, Pichia, and *Candida* while *Zygosaccharomyces bailiff* required, up to 230 ppm for inhibition in certain fruit drink at pH 3.1. Yeast can actually form $SO_2$ during juice fermentation—some *S. carlsbergensis* and *S. bayanus* strains produce up to 1,000 and 500 ppm, respectively. Molds such as *Botrytis* can be controlled on grapes by periodic gassing with $SO_2$ and bisulfite can be used to destroy aflatoxins. Both aflatoxins $B_1$ and $B_2$ can be reduced in corn.

Sodium bisulfite was found to be comparable to propionic acid in its antimicrobial activity in corn containing up to 40% moisture. Although the actual mechanisms of action of $SO_2$ is not known, several possibilities have been suggested, each supported by some experimental evidence. One suggestion is that the undissociated sulfurous acid or molecular $SO_2$ is responsible for the antimicrobial activity. Its greater effectiveness at low pH tends to support this. Vas and Ingram suggested the lowering of pH of certain foods by addition of acid as a means of obtaining greater preservation with $SO_2$. It as been suggested that the antimicrobial action is due to the strong reducing power that allows these compounds to reduce oxygen tension to a point below that at which aerobic organisms can grow, or by direct action upon some enzyme system. $SO_2$ is also thought to be an enzyme poison, inhibiting growth of microorganisms by inhibiting essential enzymes. Its use in the drying of foods to inhibit enzymatic browning is based upon this assumption. Since the sulfites are known to act on disulfide bond, it may be presumed that certain essential enzymes are affected and that inhibition ensues. The sulfites do not inhibit cellular transport.

**Propionates**

Propionic acid is a three-carbon organic acid with the following structure $CH_3CH_2COOH$. This acid and its calcium and sodium salts are permitted in breads, cakes; certain cheeses and other foods primarily as a mold inhibitor. Propionic acid is employed also as a "rope" inhibitor in bread dough. The tendency toward dissociation is low with .this compound and its salts, and these compounds are consequently active in lowacid foods. They tend to be highly specific against molds, with the inhibitory action being primarily furigistatic rather than fungicidal. With respect to the antimicrobial mode of action of propionates, they act in a manner similar to that of benzoate and sorbate. The pK of propionate is 4.87 and at a pH of 4.00, 88% of the compound is undissociated, while at a pH of 6.0, only 6.7% remains undissociated. The undissociated molecule of this lipophilic acid is necessary for its antimicrobial activity. The mode of action of propionic acid is noted above with benzoic acid.

**Nitrites and Nitrates**

Sodium nitrate ($NaNO_3$) and sodium nitrite ($NaNO_2$) are used in curing formulate for meats since they stabilize red meat colour, inhibit some spoilage and food poisoning organisms, and contribute to flavour development. The role of $NO_2$ in cured meat flavour has been reviewed by Gray and Pearson. $NO_2$ has been shown to disappear both on heating and storage. It should be recalled that many bacteria and capable of utilizing nitrate as an electron acceptor and in the process effect its reduction to nitrite. The nitrite ion is by far the more important of the two in preserved meats. This ion is highly reactive and is capable of serving both as a reducing and an oxidizing agent. In an acid environment, it ionizes to yield nitrous acid. The latter further decomposes to yield nitric oxide, which is the important product from the standpoint of colour fixation in cured meats. Ascorbate or erythrobate acts also to reduce $NO_2$ to NO. Nitric oxide reacts with myoglobin under reducing conditions to produce the desirable red pigment *nitrosomyoglobin*, as shown in the following.

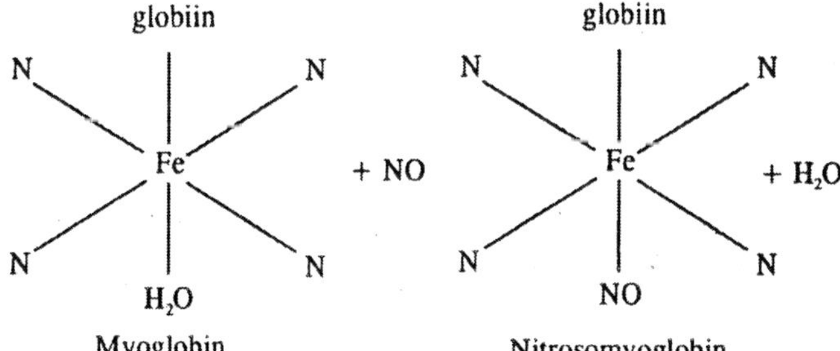


When the meat pigment exists in the form of *oxymyoglobin*, as would be the case for comminuted meats, this compound is first oxidized to *metmyoglobin*. Upon the reduction of the latter, nitric oxide reacts to yield nitrosomyoglobin. Since nitric oxide is known to be capable" of reacting with other porphyrincontaining compounds such as catalase, peroxidases, cytochromes, and others, it is conceivable that some of the antibacterial effects of nitrites against aerobes may be due to this action. It has been shown that the antibacterial effect of

$NO_2$ increases pH is lowered within the acid range, this effect is accompanied by an overall increase in the undissociated $HNO_2$.

## Organisms Affected

Although the single microorganism of greatest concentration relative to nitrite inhibition is *C. botulinum*, the compound has been evaluated as an antimicrobial for other organisms. During the late 1940s it was evaluated as a fish preservative and found to be somewhat effective but generally only at low pH. It is effective against *S. aureus* at high concentrations and, again, the effectiveness increases as pH is lowered. The compound is generally ineffective against Enterobacteriaceae, including the salmonellae, and against the lactic acid bacteria, although some effects are noted in cured and in vacuumpackaged meats and are probably caused by the interaction of nitrite with other environmental parameters rather than to nitrite alone. Nitrite is added to cheeses in some countries to control gassiness caused by *Clostridium* butyricum and *C. tyrobutyricum*. It is effective against other clostridia including *C. sporogenes* and *C. perfringens*, which are often employed in laboratory studies to assess potential antibotulinal effects not only of nitrites but of other inhibitors that might have value as nitrite adjuncts of sparing agents.

## Perigo Factor

The almost total absence of botulism in cured, canned, and vacuum packed meats and fish products led some investigators in the mid-1960s to seek reasons as to why meat products that contained viable endospores did not become toxic. Employing culture medium, it was shown in 1967 that about ten times more nitrite was needed. to inhibit clostridia if it were added after instead of before the medium was autoclaved. It was concluded that the heating of the medium with nitrite produced a substance or agent about ten times more inhibitory than nitrite alone. This agent is referred to as the Perigo factor. The existence of this factor or effect has been confirmed by some and questioned by others.

While the Perigo factor may be questionable in cured and perishable cured meats, the evidence for an inhibitory factor in culture media involving nitrite, iron, and—SH groups is more conclusive. The factor does not develop in all culture media and heating to at least 100°C is necessary for its development, although some activity develops in meats when heated to as low as 70°C. The Perigo factor is dialyzable from some culture media and meat suspensions but not from other media. It is not found in filterrilized solutions of the same medium with nitrite added. It has been shown that if meat is added to a medium containing the Perigo factor, the inhibitory activity is lost. For this reason, some Canadian workers call the inhibitor that is formed in meat the "Perigo-type factor". It is this inhibitory or antibotulinal effect that results from the heat processing or smoking of certain meat and fish products containing nitrite that warrants the continued use of nitrite in such products.

The antibotulinal activity of nitrite in cured meats is of greater public health importance than the facts of colour and flavour development. For the latter, initial nitrite levels as low as 15 to 50 ppm have been reported to be adequate for various meat products including Thuringer sausage. Nitrite levels of 100 ppm or more have been found to make for maximum flavour and appearance in fermented sausages. The antibotulinal effect requires at least 120 ppm for bacon comminuted cured ham, and canned, shelfstable luncheon meat. Many of these canned products are given a low heat process (F° of 0.1–0.6).

**Interaction with Cure Ingredients and Other Factors**

The interplay of all ingredients and factors involved in heatprocessed, cured meat on antibotulinal activity was noted over 20 years ago by Riemann and several other investigators have pointed out that curing salts in semipreserved meats are more effective in inhibiting heat-injured spores than noninjured. With brine and pH alone, higher concentrations of the former are required for inhibition as pH increases, and Change *et al.* suggested that the inhibitory effect of salt in shelf-stable canned meats against heat-injured spores may be more important than the Perigo-type factor.

With smoked salmon inoculated with $10^2$ spore/g of *C. botulinum* types A and E and stored in $O_2$ impermeable film, 3.8 and 6.1% waterp-hase NaCl alone inhibited toxin production in 7 days by types E and A, respectively. With 100 ppm or more of $NO_2$ only 2.5% NaCl was required for inhibition of toxin production by type E, and for type A 3.5% NaCl + 150 ppm $NaNO_2$ was inhibitory. With longer incubations or larger spore incoula, more NaCl or $NaNO_2$ is needed. The interplay of NaCl, $NaNO_2$, $NaNO_3$, isoascorbate, polyphosphate, thermal process temperatures, and temperature/ time of storage on spore outgrowth and germination in pork slurries has been studied extensively by Robert *et al.* who found that significant reductions in toxin production could be achieved by increasing the individual factors noted. It is well known that low pH is antagonistic to growth and toxin production by *C. botulinum* whether the acidity results from added acids or the growth of lactic acid bacteria. When 0.9% sucrose was added to bacon along with Lactobacillus *plantarum*, only one of forty-nine samples became fifty of fiftytwo samples became toxic in 2 weeks. When 40 ppm nitrite was used alone. forty-seven of fifty samples became toxic after 2 weeks but when 40 ppm nitrite was accompanied by 0.9% sucrose and an inoculum of *L. plantarum*, non of thirty became toxic. While this was most likely a direct pH effect, other factors may have been involved.

In more recent studies, bacon was prepared with 40 to 80 ppm $NaNO_2$ + 0.7% sucrose followed by inoculation with *Pediococcus acidilactici*. When inoculated with *C. botulinum* types A and B spores vacuum-packaged, and incubated up to 56 days at 27°C, the bacon was found to have greater antibotulinal properties than control bacon, prepared with 120 ppm $NaNO_2$ but not sucrose or lactic inoculum. Bacon prepared by the above formulatipn, called the Wisconsin process, was preferred by a sensory panel to process employes' 550 ppm of sodium ascorbate or sodium erythrobate, as does the' conventional process.

**Nitrosamines**

When nitrite reacts with secondary amines, *nitrosamines* are formed, and many are known to be carcinogenic. The generalized way in' which nitrosamines may form is as follows:

$$R_2NH_2 + HONO \xrightarrow{H^+} R_2N - NO + H_2O$$

The amine dimethylamine reacts with nitrite to form Nnitrosodi-methylamine. In addition to secondary amines, tertiary amines and quarternary ammonium compounds also yield nitrosamines with nitrite under acidic conditions. Nitrosamines have been found in cured meat and fish products at low levels. It has been shown that lactobacilli, group D streptococci, clostridia, and other bacteria will nitrosate secondary amines with nitrite

$$\begin{matrix} H_3C \\ \\ H_3C \end{matrix} \quad N-H + NO_2 \xrightarrow[H_3C]{H_3C} \quad N\ N=O$$

at neutral pH values. The fact that nitrosation occurred at near neutral pH values was taken to indicate that the process was enzymatic, although no cell-free enzyme was obtained. Several species of streptococci including *S. faecalis*, *S. faecium*, and *S. lactis* have been shown to be capable of forming nitrosamines, but the other lactic acid bacteria and pseudo-monads tested did not. These investigators found no evidence for an Chinese salted marine fish (previously shown to contain nitrosamines) produced nitrosamines when incoulated into salted fish homogenates containing 40 ppm of nitrate and 5 ppm nitrite.

**Nitrite-sorbate and other Nitrite Combinations**

In an effort to reduce the potential hazard of N-nitrosamine formation in bacon, the USDA in 1978 reduced the input $NO_2$, level for bacon to 120 ppm and set a 10 ppb maximum level for nitrosamines. While 120 ppm nitrite along with 550 ppm sodium ascorbate or sodium erythrobate is adequate to reduce the botulism hazard, it is desirable to reduce nitrite levels even further if protection against botulinal toxin production can be achieved. To this end a proposal to allow the use of 40 ppm nitrite in combination with 0.26% potassium sorbate. for bacon was made in 1978 but rescinded a year later when taste panel studies revealed undesirable effects.

Meanwhile, many groups of researchers have shown that 0.26% sorbate in combination with 40 or 80 ppm nitrite is effective in preventing botulinal toxin production. In an early study of the efficacy of 40 ppm nitrite+sorbate to prevent or delay botulinal toxin production in commercialtype bacon, Ivey *et al.* used an inoculum of 1,100 types A and B sores/g and incubated the product at 27°C for up to 110 days.

Time form the appearance of toxic sample when neither rite nor sorbate was used was 19 days. With 40 ppm nitrite and no sorbate, toxic samples appeared in 27 days, and for samples containing 40 ppm nitrite + 0.26% sorbate or no nitrite and 0.26% sorbate, > 110 days were required for toxic samples. This reduced nitrite level resulted in lower levels of nitrosopyrrolidine in cooked bacon. Somewhat different findings were reported by Sofos *et al.* with 80 ppm nitrite being required for the absence of toxigenic samples after 60 days. In addition to its inhibitory effects on *C. botulinum*, sorbate slows the depletion of nitrite during storage.

The effect of isoacorbate is to enhance nitrite inhibition by sequestering iron, although under some conditions it may reduce nitrite efficiency by causing a more rapid depletion of residual nitrite. EDTA at 500 ppm appears to be even more effective than erythrobate in potent-iating the nitrite effect, but only limited studies have been reported. Another chelate, 8-hydroxyquinoline, has been evaluated as a nitritesparing agent. When 200 ppm were combined with 40 ppm nitrite, a *C. botulinum* spore mixture of types A and B strains was inhibited for 60 days at 27°C in comminuted pork. In an evaluation of the interaction of nitrite and sorbate, the relative effectiveness of the combination has been shown to be dependent upon other cure ingredients and product parameters. Employing a liver-veal agar medium at pH 5.8–6.0, the germination rate of *C. botulinum* type E spores decreased to nearly zero with 1.0, 1.5 or 2.0% sorbate; but with the same concentrations at pH 7.0–7.2, germination and outgrowth of abnormally shaped cells occurred. When 500 ppm nitrite was added to the higher-pH medium along with sorbate, cell lysis was enhanced.

These investigators also found that 500 ppm linoleic acid alone at the higher pH prevented emergence and elongation of spores. Potassium sorbate significantly decreased toxin production by types A and B spores in pork slurries when NaCI was increased or pH and storage temperature were reduced. For chicken frankfurters, a sorbatebetalains mixture was found to be as effective as a conventional nitrite system for inhibiting *C. perfringens* growth.

**Mode of Action**

It appears that nitrite inhibits *C. botulinum* by interferring with iron-sulfur enzymes such a ferrodoxin and thus preventing the synthesis of ATP from pyruvate. The first direct finding in this regard was that of Woods *et al.* who showed that the phosphoroclastic system of *C. sporogenes* is inhibited by nitric oxide, and later that the same occur in *C. botulinum*, resulting in the accumulation of pyruvic acid in the medium.

**Table 4.2. Effect of nitrite and sorbate or toxin production in bacon incoulated with *C. botulinum* types A and B spores and held up to 60 days at 27°C.**

| *Treatment* | *Percent Toxigenic* |
|---|---|
| Control (no $NO_2$, no sorbate) | 90.0 |
| 0.26% sorbate, no $NaNO_2$ | 58.8 |
| 0.26% sorbate+40 ppm $NaNO_2$ | 22.0 |
| 0.26% sorbate+8p ppm $NaNO_2$ | 0.0 |
| No sorbate, 120 ppm $NaNO_2$ | 0.4 |

The phosphoroclastic reaction involves the breakdown of pyruvate with inorganic phosphate and coenzyme A to yield acetyl phosphate. In the presence of ADP, ATP is synthesized from acetyl phosphate with acetate as the other product. In the breakdown of pyruvate, electrons are transferred first to ferredoxin, and from ferredoxin to $H^+$ to form $H_2$ in a reaction catalyzed by hydrogenase. Ferrodoxin and hydrogenase are ironsulfur (nonheme) proteins or enzymes. Following the work of Woods and Wood, the next most significant finding was that of Reddy *et al.* who subjected extracts of nitriteascorbate-treated *C. botulinum* to electron spin resonance and found that nitric oxide reacted with iron-sulfur complexes to form iron-nitrosyl complexes.

The presence of the latter results in the destruction of iron-sulfur enzymes such as ferrodoxin. The resistance of the lactic acid bacteria to nitrite inhibition is well known, but the just now clear these organisms lack ferrodoxin. The clostridia contain both ferrodoxin and hydrogenase, which function in electron transport in the anaerobic breakdown of pyruvate to yield ATP, $H_2$ and CO The ferrodoxin in clostridia has a molecular weight of 6,000 and contains 8 Fe atoms/mole and 8-labile sulfide atoms/mole. Although the first definitive experimental finding was reported in 1981 as noted above, work pointed to iron-sulfur enzymes as the probable nitrite targets.

Among the first were O´ Leary and Solberg, who showed that at 91% decrease occurred in the concentration of free-SH groups of soluble cellular compounds of *C. perfringens* inhibited by nitrite. Two years later. Tompkin *et al.* offered the hypothesis that nitric oxide reacted with iron in the vegetative cells of *C. botulinum*, perhaps the iron in ferrodoxin. The inhibition by nitrite of active transport and electron transport was noted by several investigators. and these effects are consistent with nitrite inhibition of nonneme enzymes such as ferrodoxin and hydrogenase. The enhancement of inhibition in the presence of sequestering agents may be due to the reaction of sequestrants to substrate iron: more nitrite becomes available for nitric oxide production and reaction with microorganisms.

**Summary of Nitrite Effects**

The following summary of the overall role and effects of nitrite in cured meats emphasizes the antibotulinal activities. When added to processed meats such as wieners, bacon, smoked fish, and canned cured meats followed by substerilizing heat treatments, nitrite has definite antibotulinal effects. It also forms desirable product colour and enhances flavour in cured meat products. The antibotulinal effect consists of inhibition of vegetative cell growth and the prevention of germination and growth of spores that survive heat processing or smoking during post-processing storage. Clostridia other than *C. botulinum* are affected in a similar manner.

While low initial levels of nitrite are adequate for colour and flavour development, considerably higher levels are necessary for the antimicrobial effects. When nitrite is heated in certain laboratory media, an antibotulinal factor or inhibitor is formed, the exact identity of which os not yet known.

The inhibitory factor is the Perigo effect/factor or Perigo inhibitor. It does not form in filtersterilized media. It develops in canned meats only when nitrite is present during heating. The initial level o itrite is more important to antibotulinal activity than the residual level. Once formed, the Perigo factor is not affected greatly by pH changes. Measurable preheating levels of nitrite decrease considerably during heating in meats and during postprocessing storage, more at higher storage temperatures than at lower.

The antibotulinal activity of nitrite is interdependent with pH, salt content, temperature of incubation, and numbers of botulinal spores. Heat-injured spores are more susceptible to inhibition than uninjured. Nitrite is more effective under Eh-than under Eh+conditions. Nitrite does not decrease the heat resistance of spores. It is not affected by ascorbate in its antibotulinal actions but does act synergistically with ascorbate in pigment formation.

Lactic acid bacteria are relatively resistant to nitrite. Endospores remain viable in the presence of the antibotulinal effect and will germinate when transferred to nitrite-free media. Nitrite has a pK of 329 and consequently exists as undissociated nitrous acid at low pH values. The maximum undissociated state and consequent greatest antibacterial activity of nitrous acid are between pH 4.5 and 5.5. With respect to its depletion or disappearance in ham, Nordin found the rate to be proportional to its concentration and to be exponentially. related to both temperature and pH. The depletion rate doubled for every 12.2°C increase in temperature or 0.86 pH unit decrease, and was not affected by heat denaturation of the ham. These relationships did not apply at room temperature unless the product was first heat treated, suggesting that viable organisms aided in its depletion. It appears that the antibotulinal activity of nitrite is due to its inhibition of nonheme, iron-sulfur enzymes.

### Flavouring Agents

Of the many agents used to impart aromas and flavours to foods, some possess definite antimicrobial effects. In general, flavour compounds tend to be more antifungal than antibacterial. The nonlactic, gram-positive bacteria are the most sensitive and the lactic acid bacteria are rather resistant. The essential oils and spices have received the most attention by food microbiologists, while the aroma compounds have been studied more for their use in cosmetics and soaps. Of twenty-one flavouring compounds examined in one study, about one-half had *minimal inhibitory concentrations* (MIC) of 1,000 ppm or less against either bacteria or fungi. All were pH sensitive, with inhibition increasing as pH and temperature of incubation decreased. One of the most effective favouring agents is diacetyl, which imparts the aroma of butter. It is somewhat unique in being more effective against gram-negative bacteria and fungi than against gram-positive bacteria. In plate count agar at pH 6.0 and incubation at 30°C, all but one of twentyfive gram-negative bacteria and fifteen of sixteen yeasts and molds were inhibited. by 300 ppm. At pH 6.0 and incubation at 5°C in nutrient broth<.10 ppm inhibited *P. fluorescens*, *P. geniculata*, and *S. faecalis*, while under the same conditions except with incubation at 30°C, about 240 ppm were required to inhibit these and other organisms.

It appears that diacetyl antagonizes arginine utilization by reacting with argininebinding proteins of gram-negative bacteria: The greater resistance of gram-positive bacteria appears to be due to their lack of similar periplasmic binding proteins and their possession of larger amino add pools. Another flavour compound that imparts the aroma of butter is 2,3- pentanedione, and it has been found to be inhibitory to a limited number of gram-positive bacteria and fungi at 500 ppm or less. The agent 1-carvone imparts spearmintlike and the agent d-carvone imparts carawaylike aromas, and both are antimicrobial with the 1-isomer being more effective than the disomer, while both are more effective against fungi than bacteria at 1,000 ppm or less. Phenylacetaldehyde impart a hyacinthlike aroma, and has been shown to be inhibitory to *S. aureus* at 100 ppm and *Candida albicans* at 500 ppm. Menthol, which imparts a peppermintlike aroma, was found to inhibit *S. aureus* at 32 ppm, and *E. coli* and *C. albicans* at 500 ppm. Vanillin and ethyl vanillin are inhibitory, especially to fungi at levels< 1,000 ppm.

## Indirect Antimicrobials

The compounds and products in this section are added to foods primarily for effects other than antimicrobial and are this multifunctional food additives.

### Antioxidants

Although used in foods primarily to prevent the autooxidation of lipids, the phenolic antioxidants have been shown to possess antimicrobial activity against a wide range of microorganisms including some viruses, mycoplasmas, and protozoa. These compounds have been evaluated extensively as nitrite-sparing agents in processed meats and in combination with other inhibitors, and several excellent review have been made. *Butylated hydroxyanisole* (BHA), *butylated hydroxytoluene* (BHT), and *t-butylhydroxyquinoline* (TBHQ) are inhibitory to gram-positive and gram-negative bacteria as well as to yeasts and molds at concentrations ranging from about 10 to 1,000 ppm depending upon substrate. In general, higher concentrations are required to. inhibit in foods than in culture media, especially in high-fat foods. BHA was about fifty times less effective against Bacillus spp. in strained chicken than in nutrient broth.

**Table 4.3. Some indirectly antimicrobial chemicals used in foods.**

| *Compound* | *Primary use* | *Most susceptible organisms* |
|---|---|---|
| Butylated hydroxyanisole | Antioxidant | Bacteria, some fungi |
| Butylated Hydroxytoluene | Antioxidant | Bacteria, viruses, fungi |
| t-butylhydroxyquinoline | Antioxidant | Bacteria, fungi |
| Propyl gallate (PG) | Antioxidant | Bacteria |
| Nordthydroguaiaretic acid | Antioxidant | Bacteria |
| Ethylenediamine tetra-acetic acid (EDTA) | Sequestrant/ stabilizer | Bacteria |
| Sodium citrate | Buffer/seques-trant | Bacteria |
| Lauric acid | Defoaming agent | Gram-positive bacteria |
| Monolaurin | Emulsifier | Gram-positive bacteria, yeast |
| Diacetyl | Flavouring | Gram-negative bacteria, fungi |
| d-and 1-carvone | Flavouring | Fungi, gram-positive bacteria |
| Phenylacetaldehyde | Flavouring | Fungi, gram-positive |
| Menthol | Flavouring | Bacteria, fungi |
| Vanillin, ethyl vanillin | Flavouring | Fungi |
| Spices/spice oils | Flavouring | Bacteria, fungi |

BHA, BIT, TBHQ and propyl gallate (PG) were all less effective in ground pork than in culture media. While strains of the same bacterial species may show wide variation in sensitivity to either of these antioxidants, it appears that BHA and TBHQ are more inhibitory than BHT to bacteria and fungi, while the latter is more viristatic. To prevent growth of *C. botulinum* in a prereduced. medium, 50 ppm BHA and 200 ppm BHT were required, while 200 ppm. PG were ineffective. Employing 16 gram-negative and 8 gram-positive bacteria in culture media,. Gailani and Fung found the gram positives to be mor susceptible than gram negatives to BHA, BHT, TBHQ and PG with each being more effective in nutrient agar than in BHT broth.

In nutrient agar the relative effectiveness was BHA >PG> TBHQ> BHT, while in BHI, TBHQ> PG> BHA> BHT. Food-borne pathogens such as *B. cereus*, *V. paraeaemolyticus*, salmonellae, and *S. aureus* are effectively inhibited at concentrations <500 ppm, while some are sensitive to as little as 10 ppm. The pseudomonads, especially *P. aeruginosa*, are among the most resistant bacteria. Three toxin-producing penicillia were inhibited significantly in salami by BHA, TBHQ and a combination of these two at 100 ppm, while BHT and PG were ineffective. Combinations of BHA/sorbate and BHT/monolaurin have been shown to be synergistic against *S. aureus*, and BHA/sorbate against *S. typhimurium*, BHT/TBIdQ has been shown to be synergistic against aflatoxin-producing penicillia.

**Medium-chain Fatty Acids and Esters**

Acetic, propionic, and sorbic acids are short-chain fatty acids primarily as preservatives. Medium-chain fatty acids, however, are employed primarily as surface-active or emulsifying agents. The antimicrobial activity of the mediumchain fatty acids is best known from soaps, which are salts of fatty acids. Those most commonly employed are composed of twelve to sixteen carbons. For saturated fatty acids, the most antimicrobial chain length is $C_{12}$; for monounsaturated (containing 1 double bond) $C_{10\text{-}1}$; and for polyunsaturated (containing more than one double bond) $C_{10\text{-}1}$ is the most antimicrobial. (In general, fatty acids are effective primarily against gram-positive bacteria and yeasts. While the $C_{12}$ to C16 chain lengths are the most active against bacteria, the $C_{10}$ to C12 are most active against yeasts.

Fatty acids and esters and the structure-function relationships among them have been reviewed and discussed by Kabara. Saturated aliphatic acids effective against *C. botulinums* have been evaluated by Dymicky and Trenchard. The monoesters of glycerol and the diesters of sucrose have been found to be more antimicrobiakthan the corresponding free fatty acids, and to compare favourably with sorbic acid and the parabens as antimicrobials. Monolaurin is the most effective of the glycerol monoesters, while 'sucrose dicaprylate is the most' effective of the sucrose diesters.

Monolaurin (lauricidin) has been evaluated by a large number of investigators and found to be inhibitory to a variety of gram-positive bacteria and some yeasts at 5–100 ppm. Unlike the shortchain fatty acids, which are most effective at low pH, monolaurin is effective over the range 5.0 to 8.0. Because the fatty acid and esters have a narrow range of effectiveness and GRAS substances such as EDTA, citrate, and phenolic antioxidants also have limitations as antimicrobial agents when used alone, Karara has stressed the "preservative system" approach to the control of microorganisms in foods by using combinations of chemicals to fit given food systems and preservation needs.

By this approach, a preservative system might consist of three compounds-monolaurin/EDTA/BHA, for example. While EDTA possesses little antimicrobial activity by itself, it renders gram-negative bacteria more susceptible by rupturing the outer membrane and thus potentiating the effect of fatty acids or fatty acid esters. An antioxidant such as BHA would exert effects against bacteria and molds and serve as an antioxidant at the same time. By use of such a system, the development of resistant strains could be minimized and the pH of a food could become less important relative to the effectiveness of the inhibitory system.

**Acetic and Lactic Acids**

These two organic acids are among the most widely employed as preservatives. In most instances, their origin to the subject foods is due to their production within the food by lactic acid bacteria. Products such as pickles, sauerkraut, and fermented milks, among others, are created by the fermentative activities by various lactic acid bacteria, which produce acetic, lactic, and other acids. The antimicrobial effect of organic acids such as propionic and lactic is due both to the depression of pH below the growth range and metabolic inhibition by the undissociated acid molecules. In determining the quantity of organic acids in. foods, *titratable acidity* is of more value than pH alone, since the latter is. a measure of hydrogen-ion concentration and organic acids do not ionize completely. In measuring titratable acidity, the amount of acid that is Capable of reacting with a known amount of base is determined. The fitratable acidity of products. such as sauerkraut is a better indicator of the amount of acidity present than pH.

**Antibiotics**

While no antibiotic is legally permissibly as a food additive in the United States at the present time, two are approved for food use in many other countries (nisin and natamycin), and three others (tetracyclines, subtilin, and tylosin) have been studied and found effective for various food applications. The early history, efficacy, and applications of most were reviewed in 1966, and all have been reviewed and discussed more recently. Detailed reviews on nisin have been provided by Hurst and Lipinska. Three antibiotics have been investigated extensively as heat adjuncts for canned foods (subtilin, tylosin, and nisin). Nisin, however, in used most widely in cheeses. Chlortetracycline and oxytetracycline were dely studied for their application to fresh foods while natamycin is employed as a food fungistat.

While in general the use of chemical preservatives in foods is not popular among many consumers, the idea of employing antibiotics is even less popular. Some risks may be anticipated from the use of any food additive, but the risks should not outweigh the benefits overall. The general view in the United States at the present time is that the benefits to be gained by using antibiotics in foods do not outweigh the risks, some of which are known and some of which are presumed. Some fifteen considerations on the use of antibiotics as food preservatives were noted by Ingram *et al.*, and several of the key ones are summarized below:

1. The antibiotic agent should kill, not inhibit the flora, and should ideally decompose into innocuous products, or be destroyed on cooking for products that require cooking.
2. The antibiotic should. not be inactivated by food components or products of microbial metabolism.
3. The antibiotic should not readily stimulate the appearance of resistant strains.
4. The antibiotic should not be used in foods if used therapeutically or as an animal feed additive.

If may be noted from the summary comparison in Table that the tetracyclines are used both clinically and as feed additives while tylosion is used .in animal feeds and only in the treatment of some poultry diseases. Neither nisin nor subtilin issued medically or in animal feeds, and while nisin is used in many countries, subtilin is not.

## Spices and Essential Oils

While used primarily as flavouring and seasoning agents in foods, many spices possess significant antimicrobial activity. In all instances, antimicrobial activity is due to specific chemical or essential oils, some or which are noted. The search for nitritesparing agents generated new interest in spices and spice extracts in the late 1970s. It would be difficult to predict what antimicrobial eats if any are derived from spices as they are used in foods, for the quantities employed differ widely depending upon taste and the relative effectiveness varies depending upon product composition. Because of the varying concentrations of the antimicrobial constituents in different spices, and because many studies have been conducted employing them on a dry weight basis, it is difficult to ascertain the MIC of given spices against specific organisms. Another reason for conflicting results by different investigators is the assay method employed.

In general, MIC values are obtained when highly volatile compounds are evaluated on the surface of plating media than when they are tested in pour plates or broth. When eugenol was evaluated by surface plating onto PCA at pH 6, only nine of fourteen gram-negative and twelve of twenty gram-positive bacteria (including eight lactics) were inhibited by 493 ppm, while in nutrient broth at the same pH, MICs of 32 and 63 were obtained for *Torulopsis candida* and *Aspergillus nigar*, and *S. aureus* and *E. coli*, respectively. It has been noted that spice extracts are less inhibitory in media than spices, and this is probably due to a slower release of volatiles by the latter.

In spite of the difficulties of comparing results from study to study, the antimicrobial activity of spices is unquestioned and a large number of investigators have show the effectiveness of at least twenty different spices or their extracts against most food-poisoning organisms including mycotoxigenic fungi. In general, spices are less effective in foods than in culture media, and gram-positive bacteria are more sensitive than gram negatives, with the lactic acid bacteria-being the most resistant among gram positives. While results concerning them are debatable, the fungi appear to be in general more sensitive than gram-negative bacteria. Some gram negatives, however, are highly sensitive, Antimicrobial substances vary in content from the allicin of garlic to eugenol in cloves (16-18%).

When whole spices are employed, MIC values range from 1–5% for sensitive organisms. Sage and Rosemary are among the most antimicrobial as reported by various researchers, and it has been reported that 03% in culture media inhibited twenty-one of twenty-four grampositive bacteria and were more effective than all spice. With respect to specific inhibitory levels of extracts and essential oils, Huhtanen made ethanol extracts of thrity-three spices, tested them in broth against *C. botulinum*, and found that achiote and mace extracts produced an MIC of 31 ppm and were the most effective of the thirty-three. Next most effective were nutmeg, bay leaf, and white and black peppers with MICs of 125 ppm. Beuchat found that 100 ppm were cidal to *V. parahaemolyticus* in broth. Growth and aflatoxin production by *A. parasiticus* in broth were inhibited by 200-300 ppm of cinnamon and clove oils, by 150 ppm cinnamic aldehyde, and by 125 ppm eugenol. The mechanisms by which spices inhibit microorganisms are unclear and may be presumed to be different for unrelated groups of spices. That the mechanism for oregano, rosemary, sage. and thyme may be similar is suggested by the finding that resistance development by some lactic acid bacteria to one was accompanied by resistance to the other three.

**Nisin**

This is a polypeptide antibiotic structurally related to subtilin, but unlike subtilin it does not contain tryptophane residues. While the C-terminal amino acids are similar, the N-terminals are not. The first food use of nisin was by Hirsch et al. to prevent the spoilage of Swiss cheese by *Clostridium butyricum*. It is clearly the most widely used antibiotic for food preservation, with around thirty-nine countries permitting its use in. foods to varying degrees. It is not permitted in foods in the United States and Canada. Among some of its desirable properties as a food preservative are the following (1) it is nontoxic, (2) it is produced naturally by *Streptococcus lactic* strains, (3) it is heat stable and has excellent storage stability, (4) it is destroyed by digestive enzymes, it does not contribute to off-flavours or off-odors, and it has a narrow spectrum of antimicrobial activity.

The compound is effective against grampositive bacteria, primarily sporeformers, and ineffective against fungi and gram-negative bacteria. *Enterococcus (Streptococcus) faecalis* is one of the most resistant gram positives. A large amount of research has been carried out with nisin as a heat adjunct in canned foods, or as an inhibitor of heatshocked spores of *Bacillus* and *Clostridium* strains, and the MIC for preventing outgrowth of germinating spores ranges widely from 3 to > 5,000 ml or < 1 to > 125 ppm (1 pg of pure nisin is about 40 IU or RU-Reading unit). Depending upon the country and the particular food product, typical usable levels are in the range of about 2.5 to 100 ppm, although some countries do not impose concentration limits. A conventional heat process for low-acid canned foods requires an $F_a$ treatment of 6–8 to inactivate the endospores of both *C. botulinum* and spoilage organisms.

By adding nisin the heat process can be reduced to an $F_a$ of 3 (to inactivate *C. botulinum* spores) resulting in increased product quality of low-acid canned foods. While the low-heat treatment will not destroy the endospores of spoilage organisms, nisin prevents their germination

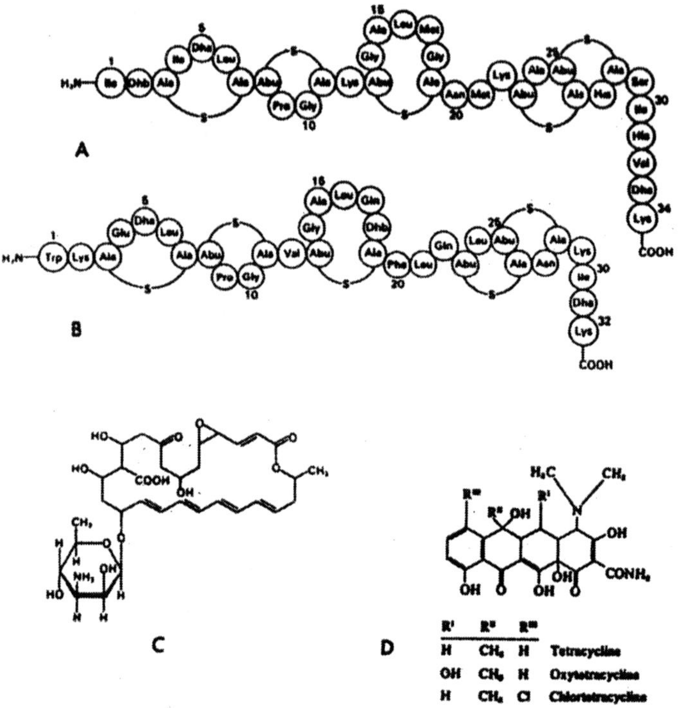

*Fig. 4.2. Structural formulae of nisin (A), subtilin (B), natamycin (C), and the tetracyclines (D).*

by acting early in the endospore germination cycle. In addition to its use in certain canned foods, nisin is most often employed in dairy products-processed cheeses, condensed milk, pasteurized milk, and so on. Some countries permit its use in processed tomato products and canned fruits and vegetables. It is most stable in acidic foods. Because of the effectiveness of nisin in preventing the outgrowth of germinating endospores of *C. botulinum* and the search to find safe substances that might replace nitrites in processed meats, this antibiotic has been studied as a possible replacement fox nitrite.

While some studies showed encouraging results employing *C. sporogenes* and other. nonpathogenic organisms, a recent study employing *C. botulinum* types A and B spores in pork slurries indicated the inability of nisin at concentrations up to 550 ppm in combination with 60 ppm nitrite to inhibit spore outgrowth Employed in, culture media without added nitrite, the quantity of nisin required for 50% inhibition of *C. botulinum* type E spores was 1–2 ppm; 10–20 ppm for type B;

and 20–40 ppm for type A. The latter authors found that higher levels were required for inhibition in cooked meat medium than in TPYG medium and suggested that nisin was approximately equivalent to nitrite in preventing the outgrowth of *C. botulinum* spores. With respect to mode of action, nisin and subtilin may be presumed to act similarly since they are both polypeptide antibiotics with highly similar structures. They act at the same site on germinating endosperm. Some of the polypeptide antibiotics typically attack cells membranes and act possibly as surfactants or emulsifying agents on membrane lipids. These agent may be presumed to inhibit gram-positive bacteria by inhibiting cell wall murein synthesis since bacitracin (another polypeptide antibiotic that also inhibits gram-positive bacteria) is known to inhibit murein synthesis. That nisin affects murein synthesis has been shown by Reisinger *et al.* and this finding is not inconsistent with its lack of toxicity for man. A similar lack of toxicity for subtilin may be presumed.

**Natamycin**

This antibiotic is a polyene that is quite effective against yeasts and molds but not bacteria. Natamycin is the international nonproprietary name since it was isolated from *Streptomyces natalensis*. In granting the acceptance of natamycin as a food preservative, the joint FAO/WHO Expert Committee took the following into consideration: (1) it does not affect bacteria, (2) it stimulates an unusually low level of resistance among fungi, (3) it is rarely involved in cross resistance among other antifungal polyenes, and (4) DNA transfer between fungi does not occur to the extent that it does with some bacteria. It may be noted that. its use is limited as. a clinical agent, and it is not used as a feed additive. The relative effectiveness of natamycin was compared to sorbic acid and four other antifungal antibiotics by Klis *et al.* for the. inhibition of sixteen different fungi (mostly molds), and while from 100 to 1,00 ppm sorbic acid were required for inhibition, from 1 to 25 ppm natamycin were effective against the same strains in the same media.

To control fungi on strawberries and raspberries, natamycin was compared with rimocidin and nystatin, and it along with rimbocidin was effective at levels of 10–20 ppm, while 50 ppm nystatin were required for effectiveness. In controlling fungi on salami, the spraying of fresh salami with a 0.25% solution was found to be effective by one group of investigators, but another researcher was unsuccessful in his attempts to prevent surface-mold growth on Italian dry sausages when they were dipped in a 2,000-ppm solution. Natamycin spray (2 × 1,000 ppm) was as good as or slightly better than 2.5% potassium sorbate. Natamycin appears to act in the same manner as other polyene antibiotics by binding to membrane sterols and inducing distortion of selective membrane permeability. Since bacteria do not possess membrane sterols, their lack of sensitivity to this agent is thus explained. Mycoplasmae, however, do have membrane sterols, but wether this antibiotic is effective against this group is unclear.

**Subtilin**

This antibiotic was discovered and developed by scientists at the Western Regional Laboratory of the USDA, and its properties were described by Dimick et al. As noted above, it is structurally similar to nisin even though it is produced by some strains of *B. subtilis*. Like nisin, it is effective against grampositive bacteria, is stable to acid, and possesses enough heat resistance to withstand-destruction at 121°C for 30 to 60 min. Subtilin is effective in canned foods at levels of 5 to 20 ppm in preventing the outgrowth of germinating endospores, and its site of action is the same as for nisin. Like niskin, it is used neither in the treatment of human or animal infections nor as a feed additive. This antibiotic may be just as effective as raisin even though it has received little attention since the late 1950s.

**Tetracylines**

Chlortetracycline (CTC) and oxytetracycline (OTC) were approved by the FDA in 1955 and 1956, respectively, at a level of 7 ppm to control bacterial spoilage in uncooked refrigerated poultry, but these approvals were subsequently rescinded. The efficacy of this group of antibiotics in extending the shelf life of refrigerated foods was first established by Tarr and associates in Canada working with fish. Subsequent research by a large number of workers in many countries established the effectiveness of CTC and. OTC in delaying bacterial spoilage of not only fish and seafoods but poultry, red meats, vegetables, raw milk, and her foods.

CTC is generally more effective than OTC. The surface treatment of refrigerated meats with 7–10 ppm typically results in self-life extensions of at least 3–5 days and a shift in ultimate spoilage flora from gram-negative bacteria to yeasts and molds. When CTC is combined with sorbate to delay spoilage of fish, the combination has been shown to be effective for up to 14 days. Rockfish fillets dipped in a solution of 5 ppm CTC and 1% sorbate had significantly lower APCs after vaccum-package storage at 2°C after 14 days than controls.

The tetracyclines are both heat sensitive and storage labile in foods, and these factors were important in their initial acceptance for food use. They are used to treat disease in man and animals and are used also in feed supplements. The risks associated with their use as food preservatives in developed countries seem clearly to outweigh the benefits.

**Tylosin**

This antibiotic is a nonpolyene macrolide as are the clinically useful antibiotics erythromycin, oleandomycin, and others. It is more inhibitory than nisin or subtilin. Denny *et at.* were apparently the first to study its possible use in canned foods. When I ppm was added to cream-style corn containing flat-sour spores and given a "botulinal" cook, no spoilage of product occurred after 30 days with incubation at 54°C. Similar findings were made by others in the 1960s, and these have been summarized. Unlike nisin subtilin, and natamycin, tylosin is used in animal fees and also to treat some diseases of poultry. As a macrolide, it is most effective against gram-positive bacteria. In inhibits protein synthesis by associating with the 50S ribosomal subunit, and shows at least partial cross-resistance with erythromycin.

**Ethylene and Propylene Oxides**

Ethylene and propylene oxides along with ethyl and methyl formate ($HCOOC_2H_5$ and $HCOOCH_3$, respectively) are treated together in this section because of their similar actions. The structures of the oxide compounds are as follows:

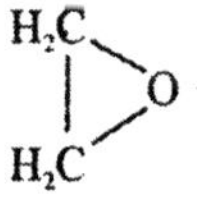

$H_2C$—O—$H_2C$ (three-membered ring)

$CH_2$—O—$CH.CH_3$ (three-membered ring)

Ethylene oxide 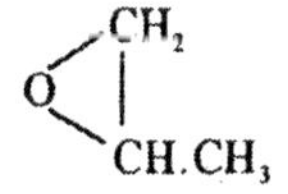 Propylene oxide

The oxides exist as gases and are employed as fumigants in the food industry. The oxides are applied to dried fruits, nuts, spices, and so forth, primarily as antifungal compounds. Ethylene oxide is an alkylating agent its antlmicrobial activity is presumed to be related to this action in the following manner. In the presence of labile H atoms, the unstable three-membered ring of ethylene oxide splits. The H atom attaches itself to the oxygen, forming a *hydroxyl* ethyl radical, $CH_2$ $CH_2$ OH, which in turn attaches itself to the position in the organic molecule left vacant by the H atom. As a result, the hydroxyl ethyl group blocks reactive groups within microbial proteins, thus resulting in inhibition. Among the groups-capable of supplying a labile H atom are the following: —COOH, —$NH_2$, —SH, and —OH. Ethylene oxide appears to affect endospores of *C. botulinum* by alkylation of guanine and adenine components of spore DNA. Ethylene oxide is. used as a gaseous sterilant for flexible and semirigid containers for packaging aseptically processed foods. All of the gas dissipates from the containers following their removal from treatment chambers. With respect to its action on microorganisms, it is not much more effective against vegetative cells than is against endospores.

**Antifungal Agents for Fruits**

Some compounds are applied to fruits after harvest to control fungi, primarily molds. *Benomyl* is applied uniformly over the entire surface of fruits, examples of which are noted in the table. It is applied at concentrations of 0.5—1.0g/liter. It can penetrate the surface of some vegetables, and is used worldwide to control crown rot and anthracnose of bananas, and stem-end rots of citrus fruits. It is more effective than *thibendazole* and penetrates with greater ease.

**Table 4.4. Some chemical agents employed to control fungal spoilage of fresh fruits**

| *Compound* | *Fruits* |
|---|---|
| Thiabensdazole | Apples, pears, citrus fruits, pineapples |
| Benomyl | Apples, pears, bananas, citrus fruits, mangoes, papayas, peaches, cherries, pine-apples |
| Biphenyl | Citrus fruits |
| $SO_2$ fumigation | Grapes |
| Sodium-a-phenylaphenate | Apples, pears, citrus fruits, pineapples |

Both benomyl and thiabendazole are effective in controlling dry rot caused by *Fusarium* spp. To prevent the spread of *Botrytis* from grape to grape, $SO_2$ is employed for long-term storage. It is applied shorly after harvest and about once a week thereafter. A typical initial treatment consists of a 20-min application of a 1% preparation, and about 0.25% in subsequent treatments. *Biphenyl* is used to control the decay of citrus fruits by peniciilia for long-distance shipments and is generally impregnated into fruit wraps or sheets between fruits layers.

**Miscellaneous Chemical Preservatives**

*Sodium diacetate* ($CH_3COONa$. $CH_3COOH$. $xH_2O$), a derivative of acetic acid, is used in bread and cakes to prevent moldiness. Organic acids such as *citric*, exert a preserving effect on foods such as soft drinks.

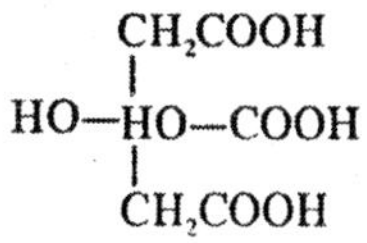

$$
\begin{array}{c} CH_2COOH \\ | \\ HO-HO-COOH \\ | \\ CH_2COOH \end{array}
$$

*Hydrogen peroxide* ($H_2O_2$) has received limited use as a food repservative. In combination with heat, it has been used in milk pasteurizaand sugar processing, but its widest use is as a sterilant for food contact surfaces of olefin polymers and polyethylene in aseptic packaging sytems.

**Table 4.5. D values for four chemical sterilants of some food-borne microorganisms.**

| *Organism* | $D_t$ | *Conc.* | *Temp-2* | *Condition* | *Reference* |
|---|---|---|---|---|---|
| | | ***Hydrogen peroxide*** | | | |
| *C. botulinum* 169B | 0.03 | 35% | 88 | | 121 |
| *B. coagulans* | 1.8 | 26% | 25 | | 122 |
| *B. stearothermophilus* | 1.5 | 26%° | 25 | | 122 |
| *B. subtilis* ATCC 95244 | 1.5 | 20% | 25 | | 116 |
| *B. subtilis A* | 7.3 | 26% | 25 | | 121 |
| | | ***Ethylene oxide*** | | | |
| *C. botulinum* 62A | 11.5 | 700mg/L | 40 | 47%R.H. | 1.06 |
| *C. botulinum* 62A | 7.4 | 700mg/L | 40 | 23% R.H. | 130 |
| C. sporogenes ATCC 7955 | 3.25 | 500mg/L | 54.4 | 40% R.H. | 68 |
| *B. coagulans* | 7.00 | 700mg/L | 40 | 33% R.H. | 7 |
| *B. coagulans* | 3.07 | 700mg/L | 60 | 33% R.H. | 7 |
| *B. stearothermophilus* ATCC 7953 | 2.63 | 500mg/L | 54.4 | 40% R.H. | 68 |
| *L. brevis* | 5.88 | 700mg/L | 30 | 33% R.H. | 7 |
| *M. radiodurans* | 3.00 | 500mg/L | 54.4 | 40% R.H. | 68 |
| | | ***Sodium hypochlorite*** | | | |
| *A. niger* conidiospores | 0.61 | 20ppmt† | 20 | pH 3.0 | 15 |
| *A. niger* conidiospores | 1.04 | 20ppmt† | 20 | pH 5.0 | 15 |
| *A. niger* conidiospores | 1.31 | 20ppmt† | 20 | pH 7.0 | 15 |
| | | ***Iodine (½$I_2$)*** | | | |
| *A, niger* conidiospores | 0.86 | 20 ppmt† | 20 | pH 3.0 | 15 |
| *A. niger* conidiospores | 1.15 | 20 ppmt† | 20 | pH 5.0 | 15 |
| *A. niger* conidiospores | 2.04 | 20 ppmt† | 20 | pH7.0 | 15 |

The D of some food-brone microorganisms are presented in Table. *Ethanol* ($C_2H_5OH$) is present in flavouring extracts and effect preservation by virtue of its desiccant and denaturant properties. *Dehydroacetic acid*, is used to

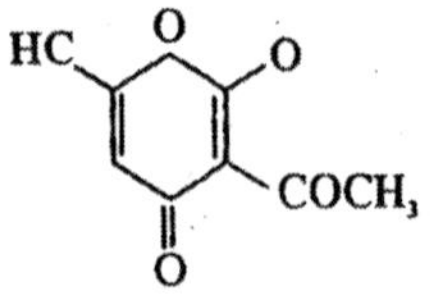

$$\text{HC}\!-\!\text{ring}(\text{O},\ =\text{O},\ \text{COCH}_3,\ =\text{O})$$

preserve squash. *DietMhylpyroearbonate* has been used in bottled wines and soft drinks as a yeasts inhibitor. It decomposes to form ethanol and $CO_2$ by either hydroylsis or alcoholysis.

Hydrolysis (reaction with Water):

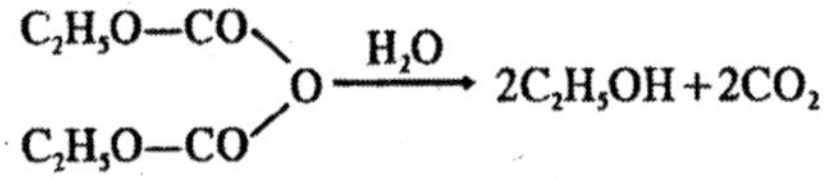

$$\begin{matrix} C_2H_5O{-}CO \\ \phantom{x}\quad\; O \\ C_2H_5O{-}CO \end{matrix} \xrightarrow{H_2O} 2C_2H_5OH + 2CO_2$$

Alcoholysis (reaction ith ethyl alcohol);

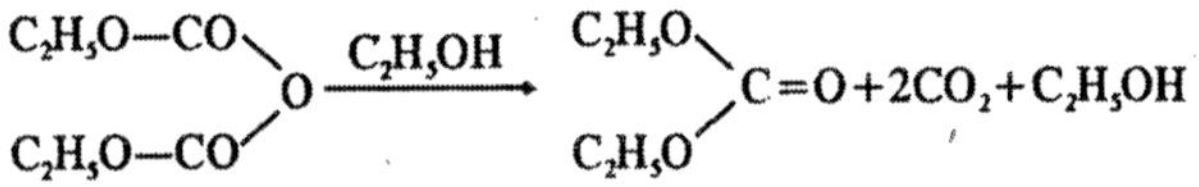

$$\begin{matrix} C_2H_5O{-}CO \\ \phantom{x}\quad\; O \\ C_2H_5O{-}CO \end{matrix} \xrightarrow{C_2H_5OH} \begin{matrix} C_2H_5O \\ \phantom{x}\quad C{=}O + 2CO_2 + C_2H_5OH \\ C_2H_5O \end{matrix}$$

*Saccharomyces cerevisiae* and conidia of *Aspergillus niger* and *Byssochlamys fulva* have been shown to be destroyed by this compound during the first'/ h of exposure, while the ascospores of *B. fulva* required 4 to 6 h for maximal destruction. Cidal concentrations for yeasts range from about 20 to 1,000 ppm depending upon species or strain. *Lactobacillus plantarum* and *Leuconostoc mesenteroides* required 24 h or longer of destruction. Sporeforming bacteria are quite resistant to this compound. Sometimes urethane is formed when this compound is used and because it is carcinogen, the use of diethylpyrocarbonate is no longer permissible in the United States.

*Wood smoke* imparts certain chemicals to smoked products that enables these products to resist microbial spoilage. One of the most important is formaldehyde ($CH_2O$), which has been known many years to possess antimicrobial properties. The compound acts as a protein denaturant by virtue of its reaction with amino groups. Also in wood smoke are aliphatic acids, alcohols, ketones, phenols, higher aldehydes, tar methanol, cresols, and other compounds, all of which may contribute to the antibacterial actions of meat smoking. Since a certain amount of heat is necessary to produce smoke, part of the shelf-stability of smoked products is due to heat destruction of surface organisms as well as to the drying that occurs. A study of the antibacterial activity of liquid smoke by Hand ford and Gibbs revealed that little activity occurred at concentrations of smoke that produced acceptable smoked flavor. Employing an agar medium containing 1:1 dilution of smoked water, these investigators found that micrococci and staphylococci were slightly more inhibited than the lactic acid bacteria. The overall combined effect of smoking and vacuum packaging results in a reduction of numbers of catalase-negative lactic acid bacteria are better able to withstand the low Eh conditions of vacuumpackaged products.

The *lactoperoxidase system* is an inhibitory system that occurs naturally in bovine milk. Itconsists of three components lactoperoxidase, thiocyanate, and $H_20_2$. All three components are required for antimicrobial effects, and the gram-negative psychrotrophs such as the pseudomonads are quite sensitive. The quantity of lactoperoxidase needed is 0.5–1.0 ppm, while bovine milk normally contains about 30 ppm. While both thiocyanate and $H_2O_2$ occur normally in milk, the quantities vary. For $H_2O_2$ about 100 U/ml are required in the inhibitory system while only 1–2 U/ml normally occurs in milk. An effective level of thiocyanate is around 0.25 mM, while in milk the quantity varies between 0.025–0.25 mM. When the lactoperoxidase system in raw milk was activated by adding thiocyanate to 025 mM along with an equimolar amount of the $H_2O_2$, shelf life was extended to 5 days compared to 48 h for controls. The system was more effective at 30 than at 4°C. The bactericidal effect increases with acidity, and the cytoplasmic membrane appears to be the cell target. In addition to the direct addition of $H_2O_2$, an exogenous source can be provided by the addition of glucose and glucose oxidase. To avoid the direct addition of glucose oxidase, this zyme has been immobilized on glass beads so that glucose is generated only in the amounts needed by the use of immobilized β-galactosidase. The lactoperoxidase system can be used to preserve raw milk in countries where refrigeration is uncommon. The addition of about 12 ppm of $SCN^-$ and 8 ppm of $H_2O_2$ should be harmless to the consumer.

### With Low Temperature

The activities of food-borne microorganisms can be slowed down and/or stopped at temperatures above freezing and generally stopped at subfreezing temperatures is the fact upon which is based the use of low temperatures to preserve foods. The reason for this is that all metabolic reactions of microorganisms are enzyme catalyzed and that the rate of enzymecatalyzed reactions is dependent on temperature. With a rise in temperature, there is an increase in reaction rate. The *temperature coefficient* ($Q_{10}$) may be generally defined as follows:

$$Q_{10} = \frac{\text{(Velocity at a given temp. } + 10)^\circ}{\text{Velocity at T}}$$

There is towfold increase in the rate of reaction. because the $Q_{10}$. for most biological systems is 1.5 to 25, so that for each 10°C rise in temperature within the suitable range. For every 10°C decrease in temperature, the reverse is true, of course. Since the. basic feature of low-temperature food preservation consists of its effect upon, spoilage organisms, most of the discussion that follows will be devoted to the effect of low temperatures on food-borne microorganisms. It should be remembered, however, that temperature is related to relative humidity (R.H.) and that subfreezing temperatures affect R.H. as well as pH and possibly other parameters of microbial growth as well.

*Psychrophiles* are the organisms that grow well at low temperatures and are referred. This term was apparently coined by Schmidt-Nielsen in 1902 for microorganisms that can grow at 0°C. The existence of such organisms in foods and especially in soils, was first noted and described by Forster in 1887. Only when, a low optimum growth temperature is implied then they should be used as suggested by Eddy. In this regard, psychrophiles should be those organisms that display maximum growth temperatures below 35C. For organisms able to grow at 5°C or less, Eddy suggested that they be called psychrotrophs. Others continue to designate as psychrophile any organism capable of growth at or around 5°C. While more and more researchers have adopted the latter terminology.

Psychrotrophic are common to a large variety of other food products and is among many of the bacteria listed in Table for meat, poultry, and the seafood products. Most psychrotrophic bacteria of importance in foods belong to the genus *Pseudomonas*, and to a lesser extent to the genera *Acinetobacter*, *Alcaligenes*, *Flavobacterium*, and others. Among the molds and yeasts, species and strains from a large number of genera are capable of growth at refrigerator temperatures. Among the low-temperaturegrowing molds are the genera *Penicillium*, *Mucor*, *Cladosporium*, *Botrytis*, and *Geotrichum*. Among yeasts, species and strains of *Debaryomyces*. *Torulopsis*, *Candida*, *Rhodotorula*, and others are known to be psychrotrophic.

Foods may be stored for preservation in at least two distinct low temperature. Chilling temperatures are those between the usual refrigerator temperatures and room temperature. These temperatures are suitable for the storage of certain vegetables and fruits, such as cucumbers, potatoes, limes, and so on. For the storage of a large number of perishable and semiperishable foods the suitable refrigerator temperatures are those between 0°-2° and 5°-7°C.

**Table 4.6. Recommended storage temperatures, relative humidity, and approximate storage life of various fresh, dried, and processed foods.**

| *Products* | *°F storage temp.* | *Percent relative humidity life* | *Approx. storage* |
|---|---|---|---|
| *Vegetables* | | | |
| Artichokes, globe | 31-32 | 90-95 | 1-2 weeks |
| Asparagus | 32 | 90—95 | 3-4 weeks |
| Beans (green or p) | 45 | 85-90 | 8-10 days |
| Beans (lima) | 32-40 | 85-90 | 10-15 days |
| Beets (bunch) | 32 | 90-95 | 10-14 days |
| Brussels sprouts | 32 | 90-95 | 3-4 weeks |
| Cabbage, late | 32 | 90-95 | 3-4 months |
| Carrots (bunch) | 32 | 90-95 | 10-14 days |
| Cauliflower | 32 | 85-90 | 2-3 weeks |
| Celery | 32 | 90-95 | 2-4 months |
| Corn, sweet | 31-32 | 85-90 | 4-8 days |
| Cucumbers | 45-50 | 90-95 | 10-14 days |
| Endive | 32 | 90-95 | 2-3 weeks |
| Lettuce | 32 | 90-95 | 3-4 weeks |
| Onions | 32 | 70-75 | 6-8 months |
| Peas, green | 32 | 85-90 | 1-2 weeks |
| Potatoes, early crop | 50-55 | 85-90 | - |
| Potatoes, sweet | 55-60 | 90-95 | 4-6 months |
| Radishes (spring, bunch) | 32 | 90-95 | 10 days |
| Rhubarb | 32 | 90-95 | 2-3 weeks |
| Rutabagas | 32 | 90-95 | 2-4 months |
| Spinach | 32 | 90-9 | 10-14 days |
| Squash, summer | 32-40 | 85-95 | 10-14 days |
| Tomatoes ripe | 32 | 85-90 | 7 days |
| *Dairy Products* | | | |
| Butter | 32-36 | 80-85 | 2 months |
| Cheese, process American, process Swiss | 40-45 | 75 | - |

| | | | |
|---|---|---|---|
| Eggs, shell | 29–31 | 85–90 | 8–9 months |
| Eggs (dried, yolk) | 35 | minimum | 6–12 months |
| *Meat, Poultry, and Fish* | | | |
| Beef, fresh | 32–34 | 88–92 | 1–6weeks |
| Hams and shoulders, fresh | 32–34 | 85–90 | 7–12 days |
| Hams and shoulders, cured | 60–65 | 50–60 | 0–3 years |
| Poultry, fresh | 32 | – | 1 week |
| Fish, fresh | 33–40 | 90–95 | 5–90 days |
| *Fruits* | | | |
| Apples | 30–32 | 85–90 | 2–7 months |
| Berries, blue | 30–32 | 85 | 3–6 weeks |
| Berries, cranberries | 36–40 | 85–90 | 1–3 months |
| Dried fruits | 32 | 50–60 | 9–12 months |
| Figs (fresh) | 28–32 | 85–90 | 4–8 weeks |
| Grapefruit | 32–50 | 85–90 | 4–8 weeks |
| Grapes (American type) | 31–32 | 85–90 | 3–8 weeks |
| Lemons | 32, 55–58 | 85–90 | 1–4 months |
| Melons (honeydew, honeyball) | 45–50 | 85–90 | 2–4 weeks |
| Melons, watermelons | 36–40 | 85–90 | 2–3 weeks |
| Oranges | 32–34 | 85–90 | 8–12 weeks |
| Peaches | 31–32 | 85–90 | 2–4 weeks |
| Strawberries, fresh | 31–32 | 85–90 | 7–10 days |

The growth of most food poisoning bacteria can be prevented at temperatures below 6°C. Those demonstrated to grow at or below 6°C are *C. botulinum* type E and nonproteolytic B and F strains; strains of *V. parahaemolyticus* and *Y. enterocolitica*; and some *A. hydrophila* strains associated with gastroenteritis. While not a food-poisoning bacterium in the traditional sense, *Listeria monocytogenes* grows well at 4°C. Below 6°C the growth of all food spoilage organisms is effectively retarted the mechanisms of this action are discussed. Temperatures at or below 18°C are freezer temperatures. Under normal circumstances, these temperatures are sufficient to prevgnt the growth of all microorganisms, but as will be seen later, some can do grow with within the freezer range, but at an extremely slow rate.

## Temperature Growth Minima of Food-borne Microorganisms

The findings of various authors who had reported the growth of microorganisms at and below–10°C. have been summarised by Michener and Elliot in their excellent review of the literature on the temperature growth mimina of microorganisms. Of the thirteen organisms reported, there were six bacteria four yeasts, and three molds. They yeasts grew at lower temperatures than the others, with one pink yeast reported to grow at -34°C and two others at -18°C. The lowest recorded temperature of growth for a bacterium is -20°C, and several have been reported to grow at about 2°C. The molds grew about -12°C. Fruit juice concentates, bacon, ice cream, and certain fruits are some foods that tend to support microbial growth at subfreezing temperatures. The reported temperature minima for fecal indicator organisms vary between-2 to 10°C.

In general, the entrococci grow better at refrigerator temperatures than do coliform bacteria. The lowest reported temperatures for growth of staphylococci and salmonellae in foods is 6.7°C. As noted above, *C. botulinum type* E has been reported to grow at a temperature lower than all other food-poisioning microorganisms; Schmidt *et al.* reported 3.3°C for this this organism in beef stow. Minimum temperature of growth and toxin production reported by many investigators for types A and B strains of *C. botuhnum* in 10°C. The reader is referred for further discussion of the temperature growth minima of fecal indicator and foodpoisoning microorganisms.

**Preparation of Food for Freezing**

Selecting, sorting, washing, blanching and packaging prior to actual freezing, are included for the preparation of vegetables for freezing. In selecting foods to be frozen, those in nay state of detectable spoilage should be rejected. Meats, poultry, seafoods, eggs, and other foods should be as fresh ass possible. Either by brief immersion of foods into water or the use of steam Blanching is achieved.

Its primary functions are as follows: (1) inactivation of enzymes that might cause undersirable changes during freezing storage, (2) enhancement or fixing of the green colour of certain vegetables, (3) reduction in the numbers of microorganisms on the foods, (4) facilitating the packing of leafy vegetables by inducing wilting, and (5) the displacement of entrapped air in the plant tissues. The products in question, size of packs and other related information is that, upon which the method of blanching depends. When water is used, it is important that bacterial spores not be allowed to build up sufficiently to contaminate foods. With respect to the reduction in numbers of microorganisms by blanching, reductions of initial microbiol loads as high as 99% have been claimed. It should be rememembered, that most vegetative bacterial cells can be destroyed at milk pasteurization temperatures (145°F for 30 min). This is especially true of most bacteria of importance in the spoilage of vegetables. To destroy microorganisms it is not the primary function of blanching, but to effect destruction of most food enzymes the amount of heat necessary is also sufficient to reduce vegetative cells significantly.

**Freezing of Food and Freezing Effects**

Quick and slow freezing are the two basic ways to achieve the freezing of foods. Quick or fast freezing is the process by which the temperature of foods is lowered to about -20°C within 30 min. This treatment may be achieved by direct immersion or indirect contact of foods with the refrigerant, and the use of airblasts of frigid air blown across the foods being frozen.

Slow freezing refers to the process whereby the desired temperature is achieved within 3 to 72 h. This is essentially the type of freezing utilized in the home freezer.

From the standpoint of overall product quality more advantages are possessed by quick freezing than slow freezing. The two methods are compared below.

| *Quick Freezing* | *Slow Freezing* |
|---|---|
| 1. Small ice crystals formed. | 1. Large ice crystals formed. |
| 2. Blocks or suppresses metabolism. | 2. Breakdown of metabolic report. |
| 3. Brief exposure to concentration of adverse constituents. | 3. Longer exposure to adverse or injurious factors. |
| 4. No adaptation to low | 4. Gradual adaptation temperatures. |
| 5. Thermal shock (tocl brutal a transition). | 5. No shock effect |
| 6. No protective effect | 6. Accumulation of concentrated solutes with beneficial effects. |
| 7. Microorganisms frozen into crystals? | |
| 8. Avoid internal metabolic imbalance. | |

The formation of small, intracellular ice crystals, with respect to crystal formation upon freezing is favoured by quick freezing while large extracellular crystals and favoured by slow freezing. Crystal growth is one of the factors that limits the freezer life of certain foods, since ice crystals grow in size and cause cell damage by disrupting membranes, cell walls, and internal structures to the point where the thawed product is quite unlike the original in texture and flavor. Upon thawing, foods frozen by the slow freezing method tend to lose more drip (drip for meats; leakage in the case of vegetables) than quickfrozen foods held for comparable periods of time.

The overall advantages of small crystal formation to frozen food quality may be viewed also from the standpoint of what takes place when a food is frozen. During the, freezing of foods, water is removed from solution and transformed into ice crystals of a variable but high degree of "purity. pH, titratable acidity, ionic strength,. viscosity, osmotic pressure, vapour pressure, freezing point, surface- and interfacial tension, and O/R potential are such properties in which changes are accompanied in addition to the freezing of foods. Some of the many complexities of this process are discussed by Fennema *et al.*

**Storage Stability of Frozen Foods**

As previously mentioned many investigators have reported, a large number of microorganisms to grow at and below 0°C. In addition to factors inherent within these organisms, their growth at and below freezing temperature is dependent upon several factors of foods, namely, nutrient content, pH, and the availability of liquid water. The $a_w$ The $a_w$ of foods may be expected to decrease as temperatures fall below the freezing point. In table the relationship between temperature and $a_w$ of water and ice is presented. For water at 0°C, $a_w$ is 1.0 but falls to about 0.8 at-20°C and to 0.62 at about-50°C. Organisms that grow at subfreezing temperatures, then, must be able to grow at the reduced $a_w$ level unless $a_w$ at levels higher than would be expected in pure water hereby making microbial growth possible even at subfreezing temperatures.

**Table 4.7. Vapour pressures of water and ice at various temperatures.**

| °C | *Liquid water* *mm.Hg* | *Ice* *mm. Hg* | *Pice* $a_w$ = *Pwater* |
|---|---|---|---|
| 0 | 4.579 | 4.579 | 1.00 |
| – 5 | 3.163 | 3.013 | 0.953 |
| – 10 | 2.149 | 1.950 | 0.907 |
| – 15 | 1.436 | 1.241 | 0.864 |
| – 20 | 0.943 | 0.776 | 0.823 |
| – 25 | 0.607 | 0.476 | 0.784 |
| – 30 | 0.383 | 0.286 | 0.75 |
| – 40 | 0.142 | 0.097 | 0.98 |
| – 50 | 0.048 | 0.030 | 0.62 |

**Table 4.8. Storage life of frozen foods-average life in "good condition" for products processed and package in a normal manner.**

| *Product* | *Months at* *-10°F* | *0°F* | *+5°F* | *+10°F* |
|---|---|---|---|---|
| Apricots | 24 | 18–24 | – | 6.8 |
| Asparagus | 16–18 | 8–12 | – | 4.6 |
| Beans, green, | 16–18 | 8–12 | 4–6 | |
| Beans, lima | 24 | 14–16 | – | 6–8 |
| Broccoli | 24 | 14–16 | 6–8 | |
| Brussels sprouts | 16–18 | 8–12 | – | 4–6 |
| Cauliflower | 24 | 14–16 | – | 6–8 |
| Corn, on cob | 12–14 | 8–10 | 4–6 | |
| Corn, cut | 36 | 24 | – | 12 |
| Carrots | 36 | 24 | – | 12 |
| Fish, fatty | 10–12 | 6–8 | – | 4 |
| Fish, lean | 14–16 | 10–12 | -6 | |
| Lobsters | 10–12 | 8–10 | – | 3–4 |
| Peaches | 24 | 18–24 | – | 6–8 |
| Peas | 24 | 14–16 | 6–8 | |
| Raspberries, sugard | 24 | 18 | – | 8–10 |
| Spinach | 24 | 18 | – | 8–10 |
| Strawberries, sliced | 24 | 18 | – | 8–10 |
| *Beef* | | | | |
| roasts, steaks | – | 12–14 | 8–10 | – |
| cubed, small pieces | – | 10–12 | 6–8 | – |
| ground | – | 8 | 4 | – |
| *Veal* | | | | |
| roasts, chops | – | 10–12 | 8–10 | – |
| thin cutlets, cubes | – | 8–10 | 4–6 | – |
| ground | – | 6 | 3 | – |
| *Lamb* | | | | |
| roasts, chop | – | 12–14 | 8–10 | – |
| cubed | – | 10–12 | 8–10 | – |
| ground | – | 8 | 4 | – |
| *Pork* | | | | |
| roasts, chops | – | 6–12 | 3–5 | – |
| ground, sausage | – | 4 | 2 | – |
| pork or ham, smoked | – | 5–7 | 3 | – |
| bacon | – | 3 | 1 | – |
| *Lard* | | | | |
| unsalted | – | 3–6 | 1–2 | – |

| | | | | |
|---|---|---|---|---|
| salted | – | 1–3 | 1/2 | – |
| *Variety meats* | | | | |
| beef or lamb liver, heart | – | 4 | 2 | – |
| veal liver, heart | – | 3 | 1 | – |
| pork liver, heart | – | 2 | 1 | – |
| tongue | – | 4 | 2 | – |
| Kidneys | – | 3 | 1 | – |
| sweet breads | – | 1 | – | – |
| brains | – | 1 | – | – |
| oxtails | – | 4 | 2 | – |
| *spiced sausage or delicatesssen* | | | | |
| meats | – | 2–3 | 1 | – |

By the addition of glycerol to culture media the same type of effect can be achieved. It should be borne in mind also that not all foods freeze at the same initial point. The initial freezing point of a given food is due in large part to the nature of its solute constituents and the relative concentration of those that have freezing-point depressing properties. If the thawed product is to retain the original flavour and texture frozen foods may not be kept indefinitely, though the metabolic activities of all microorganisms can be stopped at freezer temperature. Most frozen foods are assigned a freezer life, and suggested maximum holding periods for various foods are presented in table 4.8.

The suggested maximum holding time for frozen foods is not based upon the microbiology of such foods but upon such factors as texture, flavour, tenderness, colour, and overall nutritional quality upon thawing and subsequent cooking. Freezer burn are undergo by some foods that are improperly wrapped during freezer storage. This condition is characterized by a browing of light-coloured foods such as the skin of chicken meat. At the affected site the product is left more porous than the original as a result of browing from the loss of. moisture at the surface. The condition is irreversible and is known to-affect certain fruits, poultry, meats, and fish, both raw and cooked.

**Effect of Freezing upon Microorganisms**

It is well known that freezing is one means of preserving microbial cultures considering the effect of freezing upon those microorganisms that are unable to grow at freezing temperatures, with freeze drying being perhaps the best method known. However, freezing temperatures have been shown to effect the killing of certain microorganisms of importance in foods. Ingram has summarized the salient facts of what happens to certain microorganisms upon freezing:

1. There is sudden mortality immediately on freezing, varying with species.
2. The proportion of cells surviving immediately after freezing is nearly independent of the rate of freezing.
3. The cells that are still viable immediately after freezing die gradually when stored in the frozen state.
4. This decline in numbers is relatively rapid at temperatures just below, the freezing point, especially about -20°C, but less at lower temperatures, and it is usually slow below -20°C.

The cocci is generally more resistant than gram negative rods as bacteria differ in their capacity to survive during freezing. Of the food-poisoning bacteria, salmonellae are less resistant than *S. aureus* or vegetative cells of clostridia, while endospores and food-poisoning toxins are apparently unaffected by low temperatures. In table is presented the effect of freezing several species of Salmonella to -25.5°C and holding upto 270 days. Although a significant reduction in viable numbers occurred over the 270-day storage period with most—ties, in no instance did all cells die off. Freezing should not be regarded as a means of destroying food-borne microorganisms from the strict standpoint of food preservation.

The type of organisms that lose their viability in this state differ from strain to strain and depend upon the type of freezing employed, the nature and composition of the food in question, the length of time of freezer storage, and other factors, such as temperature of freezing. Low freezing temperatures of about -20°C are less harmful to microorganisms than the median range of temperatures such as -10°C. For example, more microorganisms are destroyed at -4°C than at -15°C or below. Temperatures below -24°C seem to have no additional effect. Cells viability has been reported to decrease by acid condition while freezing viability increases by food constituments such as egg white, sucrose, corn syrup, fish, glycerol, and undernatural meat extracts. To consider further the effects of freezing upon microorganisms, it would be of value to consider some of the events that are known to occur when cells freeze:

1. The water that freezes is the so-called free water. Upon freezing, the free water forms ice crystals. The growth of ice crystals occurs by accretion so that all of the free water of a cells might be represented by a relatively small number of ice crystals. Ice crystals are entracellular in slow freezing while intracellular in

**Table 4.9. The survival of pure cultures of enteric organisms in chicken chow mein at –25.5°C.**

| Organism | Bacterial count ($10^5$/g) after storage for (days) | | | | | | | | |
|---|---|---|---|---|---|---|---|---|---|
| | *0* | *2* | *5* | *9* | *14* | *28* | *50* | *92* | *270* |
| Salmonella *newington* | 7.5 | 56.0 | 27.0 | 21.7 | 11.1 | 11.1 | 3.2 | 5.0 | 2.2 |
| *S. typhi-nrium* | 167.0 | 245.0 | 134.0 | 118.0 | 11.0 | 95.5 | 31.0 | 90.0 | 34.0 |
| *S typhi* | 128.5 | 45.5 | 21.0 | 17.3 | 10.6 | 4.5 | 2.6 | 2.3 | 0.83 |
| *S. gallinarum* | 68.5 | 87.0 | 45.0 | 36.5 | 29.0 | 17.9 | 14.9 | 8.3 | 4.0 |
| *S anatum* | 100.0 | 79.0 | 55.0 | 52.5 | 33.5 | 29.4 | 22.6 | 16.2 | 4.2 |
| *S. paratyphi B* | 23.0 | 205.0 | 118.0 | 93.0 | 92.0 | 42.8 | 24.3 | 38.8 | 19.0 |

fast freezing. Bound water remains unfrozen. The freezing of cells depletes them of usable liquid water and thus dehydrates them.

2. Freezing results in an increase in the viscosity of cellular matte, a direct consequence of water being concentrated in the form ice crystals.
3. Freezing results in a loss of cytoplasmic gases such as $O_2$ and $CO_2$. A loss of $O_2$ to aerobic cells suppresses respiratory reaction. Also, the more diffuse state of $O_2$ may make for greater oxidative activities within the cell.
4. Freezing causes-changes in pH of cellular matter. Various auth have reported changes ranging from 0.3–2.0 pH units. Increases and decreases of pH upon freezing and thawing have been reported.
5. A concentration of cellular electrolytes is effected by freezing which in the form of ice crystals is effected as a consequence of concentration of water.
6. Freezing causes a general alteration of the colloidal state of cellular protoplasm. It should be recalled that many of the constituents of cellular protoplasm such as proteins exist in a dynamic colloidal state in living cells. A proper amount of water is necessary to the well-being of this state.
7. Freezing causes some denaturation of cellular proteins. Precisely how this effect is achieved is not clear, but it is known that some —SH groups as lipoproteins break apart from others upon freezing. The lowered water content along with the concentration of electrolytes no doubt affect this change in state of cellular proteins.
8. In some microorganisms temperature shocks is induced by freezing which more than psyehrophiles is true for thermophiles and

mesophiles. It has been shown that more cells die when the temperature. decline above freezing is sudden than when it is slow.

9. Some microbial cells such as pseudomonas spp. are caused metabolic injury by freezing. It has been reported that some bacteria have increased nutritional requirements upon thawing from the frozen state and that as much as 40% of a culture may be affected in this way.

The above should serve to illustrate the complex effects of the freezing process upon living cells such as bacteria and other microorganisms as well as upon foods. According the Mazur, the response of microorganisms to subzero temperatures appears to be largely determined by solute concentration and intracellular freezing, although there are only a few cases of clear demonstration of this conclusion. Yet not all cells but why some bacteria are killed by freezing most of bacteria in freezing some small and microscopic organisms are unable to survive examples of these include the foot-and-mouth disease virus and the causative agent of trichinosis (*Trichinella spiralis*). Protozoa are generally killed when frozen below -5 or −10°C, if protective compounds are not present.

## Effect of Thawing

The process of thawing is of great importance in the freezing survival of microorganisms. It is well established that repeated freezing and thawing will destroy bacteria by disrupting cell membranes. It is also known that the faster the thaw, the greater the number of bacterial survivors. Just why this is so is no entirely clear. If lead to the restoration of viable activity it can be -seen that thawing process becomes complicated from the changes listed above that occur during freezing. Fennema has pointed out that thawing is inherently slower than freezing and follows a pattern that is potentially more detrimental. Among the problems attendant on the thawing of specimens and products that transmit heat energy primarily by conduction are those summerized below:

1. Thawing is; inherently slower than freezing when conducted under comparable temperature differentials.
2. In practice, the maximum temperature differential permissible during thawing is much less than that which is feasible during freezing.
3. Than **freezing** potentially. more deterimental is the time temperature pattern characteristics of thawing. During thawing, the

temperatures rises rapidly to near the melting point and remains there throughout the long course of thawing, thus affording considerable opportunity for chemical reactions, recrystallization, and even microbial growth if thawing is extremely slow.

During the thawing process, as stated the microorganisms die but not upon freezing per se. Whether or not this is the case remains to be proven. As to why some organisms are able to survive freezing while others are not, Luyet has suggested that it is a question of the ability of an organism to survive dehydration and to undergo dehydration when the medium freezes. Luyet has further stated that the small size of bacteria cells permits them to undergo dehydration upon freezing. This author has stated that it might be due to the fact that bacteria do not freeze at all but merely dry with respect to survival after freeze-drying. Most frozen foods processors advise against the refreezing of foods once they have thawed. While the reasons for this are more related to the texture, flavour, and other nutritional qualities of the frozen product, the microbiology of thawed frozen foods is pertinent. Some investigators have pointed out that foods thawed from the frozen state spoil faster then similar fresh products.

The spoilage process is consequently facilitated and the invasion of surface organism is seen to be aided by the textural organisms into deeper parts of the product. Upon thawing, surface condensation of water is known to occur. There is also, at the surface, a general concentration of water-soluble substances such as amino acids, minerals, B vitamins, and possibly other nutrients, has been distroyed as a effect of freezing. Many thermophilic and some mesophilic organisms, making for less competition among the survivors upon thawing. It is conceivable that a greater relative number of psychrotrophs on thawed foods might increase the spoilage rate.

Some psychrotrophic bacteria have been reported to have $Q_{10}$ values in excess of 4.0 at refrigerator temperatures. For example P. fragi has been reported to possess a $Q_{10}$ of 4.3 at 0°C. With only a 45 degree rise in temperature organisms of this type are capable of doubling their growth rate. Whether or not frozen thawed foods do in fact spoil faster than fresh foods would depend upon a large number of factors, such as the type of freezing, the relative numbers and types of organisms on the product is held to thaw. This act should be minimized in the interest of the overall nutritional quality of the products as there are no known toxic effects associated with the refreezing of frozen and thawed foods. One effect of freezing and thawing animal tissues is the release of lysosomal enzymes consisting of cathepsins, nucleases, phosphatases, glycosidases, and others. Once released, these enzymes may act to degrade macromolecules and thus make available simpler compounds that are more readily utilized by the spoilage flora.

## High Temperature

Destructive effects on microorganisms is based on the use of high temperatures to preserve food. Above ambient any and all temperatures are meant high temperatures. With respect to food preservation, there are two temperature categories in common use pasteurization and sterilization. Pasteurization by use of heat implies either the destruction of all disease-producing organisms (for example, pasteurization of milk) or the destruction or reduction in number of spoilage organisms in certain foods, as in the pasteurization of vinegar. By heating at 145°F for 30 min, or at 161°F for 15 see (high temperature short time—HIST method) the pasteurization of milk is achieved. *Mycobacterium tuberculosis* and *Coxiella burnetti* are the most heatresistant of the non-spore-forming pathogenic organisms which are sufficiently destroyed by these treatments. Milk pasteurization temperatures are sufficient to destroy, in addition, all yeasts, molds, gram-negative bacteria, and many gram positives. Placed into one of two groups: thermodurics and thermophiles are the two groups of organisms that survive milk pasteurization.

Thermoduric organisms are those that can survive heat treatment at relatively high temperatures but do not necessarily grow at these temperatures. The nonsporeforming organisms that survive milk pasteurization generally belong to the genera *Streptococcus* and *Lactobacillus*, and sometimes to other genera. Thermophilic organisms are those that not only survive relatively high temperatures but require high temperatures for their growth and metabolic activities. The thermophiles of greatest importance in foods are contained by the genera Bacillus and clostridium.

Sterilization means the destruction of all viable organisms as may be measured by an appropriate plating or enumerating technique. Canned foods are sometimes called "commercially sterile" to signify that no viable organisms can be detected by the usual cultural methods employed, or that the number of survivors is so low as to be of no significance under the conditions of canning and storage. By reason of undesirable pH, Eh, or temperature of storage, microorganisms which may also be present is canned for that cannot grow the product. The use of ultrahigh temperatures (UHT) is a more recent development in the processing of milk and milk products. Milk so produced is a product in its own right and is to be distinguished from pasteurized milk. The primary features of the UHT treatment include: (1) its continuous nature, (2) its occurrence outside of the package necessitating aseptic storage and aseptic handling of the product downstream from the sterilizer, and (3) the very high temperatures (in the range of 140°-150°C) and the correspondingly short time (a few seconds) necessary to achieve commercial sterility. Since UHT processed milks are commercially sterile they may be stored at room temperatures for up to 8 weeks without flavour changes and also they have higher consumer acceptability than the conventionally heated pasteurized products.

**Factors that Affect Heat Resistance in Microorganisms**

Placed in physiologic saline and nutrient broth at the same pH equal numbers of bacteria are not destroyed with the same case by heat as it is well known. Some eleven factors or parameters of microorganisms and their environment have been studied for their effective on heat destruction and are presented below.

**Water**

With decreasing humidity or moisture the heat resistance of microbial cells increases. Dried microbial cells places into test tubes and then heated in a water bath are considerably more heat resistant than moist cells of the same type. Since it is well established that protein denaturation occurs at a faster rate when heated in water than in air,. it is suggested that protein denaturation is either the mechanism of death by heat or is closely associated with it. The precise manner in which water facilitates heat denaturation of proteins is not entirely clear, but it has been pointed out that the heating of wet proteins causes the formation of free-SH groups with a consequent increase in the water-binding capacity of proteins. Greater refractivity to heat in consequently conferred by a process that requires more energy is the absence of water as the presence of water allows for thermal breaking of peptide bonds.

**Fat**

In the presence of fats, there is a general increase in the heat resistance of some microorganisms. This is sometimes referred to as fat protection and is presumed to increase heat resistance by directly affecting cell moisture. On *C. botulinum* the heat protective effect of long chain fatty acids has been demonstrated by Sugiyama. It appears that the long-chain fatty acids are better protectors than short-chain acids.

**Table 4.10. The effect of the medium upon the thermal death point of *Escherichia coli*.**

| *Medium* | *Thermal death point (°C)* |
|---|---|
| Cream | 73 |
| Whole milk | 69 |
| Skim milk | 65 |
| Whey | 63 |
| Boulotin (broth) | 61 |

**Salts**

The effect of salt on the heat resistance of microorganisms is variable and dependent upon the kind of salt, concentration employed, and other factors. It has been observed that some salts have a protective effect upon microorganisms while others tend to make cells more heat sensitive. It has been suggested that some salts may decreased water activity and thereby increase heat resistance by a mechanism similar to that of drying, while others may increase water activity (for example, $Ca^{2+}$ and $Mg^{2+}$) and consequently increase sensitivity to heat. It has been shown that supplementation of the growth medium of *B. megaterium* spores with $CaCl_2$ yields spores with increased heat resistance, while the addition of L-glutamate, L-proline, or increased phosphate content decreases heat resistance.

**Carbohydrates**

An increase in the heat resistance of microorganisms suspended therein is caused by the presence of sugars in the suspending menstrum. This effect is at least in part due to the decrease in water activity that is caused by high concentrations of sugars. There is great variation, however, among sugars and alcohols relative to their effect on heat resistance for D values of *Salmonella senftenberg* 775W. Wide differences in heat sensitivity occur at identical $a_w$ values obtained by use of glycerol and sucrose. Corry found that sucrose increased the heat resistance of *S. senftenberg* more than any of four other carbohydrates tested. The following decreasing order was found for the five tested substances: sucroses > glucose > sorbitol > fructodse > glycerol.

**pH**

About pH *7.0* which is optimum pH of growth microorganisms are most resistant to heat. As pH is lowered or raised from this optimum value, there is a consequent increase in heat sensitivity.

**Table 4.11. Reported D values of *Salmonella seftenbergy* 775W as functions of various parameters of growth and other conditions.**

| Temp. (C) | D Values | Conditions |
|---|---|---|
| 61 | 1.1 min | Liquid whole egg |
| 61 | 1.19 min | Tryptose broth |
| 60 | 9.5 min | Liquid whole egg, pH ca. 5.5 |
| 60 | 9.0 min | Liquid whole egg, pH ca. 6.6 |
| 60 | 4.6 min | Liquid whole egg, pH ca. 7.4 |
| 60 | 0.36 min | Liquid whole egg, pH ca. 8.5 |
| 65.6 | 34–35.3 sec | Milk |
| 71.7 | 1.2 sec | Milk |
| 70 | 360–480 min | Milk chocolate (11) |
| 55 | 4.8 min | TSB, log phase, grown 35°C |
| 55 | 12.5 min | TSB, log phase, grown 44°C |
| 55 | 14.6 min | TSB, stationary, grown 35°C |
| 55 | 42.0 min | SB, stationary, grown 44°C |
| 57.2 | 13.5 min | $a^w$ 0.99 (4.9% glyc,), pH 6.9 |
| 57.2 | 31.5 min | $a^w$ 0.90 (33.9% glyc.), pH 6.9 |
| 57.2 | 14.5 min | $a^w$ 0.99 (15.4% sucro.), pH 6.9 |
| 57.2 | 62.0 min | $a^w$ 0.90 (58.6% sucro.), pH 6.9 |
| 60 | 0.2–6.5 | min HIB[d], pH 7.4 |
| 60 | 2.5 min | $a^w$ 0.90, HIB, glycerol |
| 60 | 75.2 min | $a^w$ 0.90, HIB, sucrose |
| 65 | 0.29 min | 0.1 M phosphate buf., pH 6.5 |
| 65 | 0.8 min | 30% sucrose |
| 65 | 43.0 min | 70% glucose |
| 65 | 2.0 min | 30% glucose |
| 65 | 17.0 min | 70% glucose |
| 65 | 0.95 min | 30% glycerol |
| 65 | 0.70 min | 70% glycerol |
| 55 | 35 min | aw 0.997, tryptone soya agar, pH 7.2 |

Advantage is taken of this fact in the heat processing of high-acid foods where considerably less heat is applied to achieve sterilization compared to foods at or near neutrality. An example of an alkaline food product that is neutralized prior to heat treatment is provided by the heat pasteurization of egg white, a practice which is not done with other foods. The pH of egg white is about 9.0. When this product is subjected to pasteurization conditions of 60–62°C for 3.5–4 min,

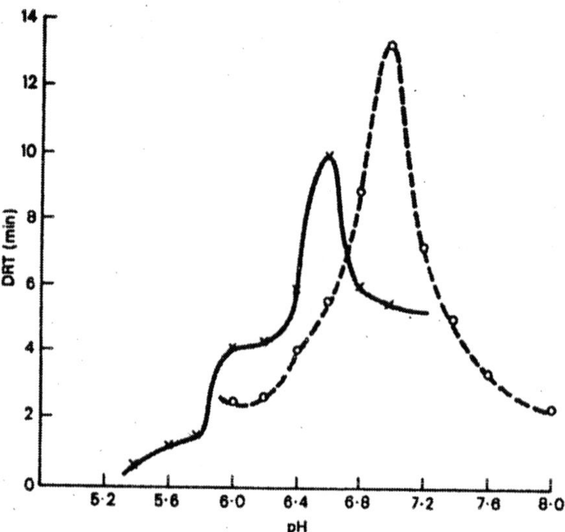


*Fig. 4.3. The effect of pH on the DRT of Streptococcus facecalis exposed to 60°C in citrate-phosphate buffer (crosses) and phosphate buffer (circle) solutions at various pH levels. DRT=decimal reduction time.*

coagulation of proteins occurs along with a marked increase in viscosity. Volume and texture of cakes made from such pasteurized egg white is affected by these changes. Cunningham and Lineweaver have reported that egg white may be pasteurized the same as whole egg it the pH is reduced to about 7.0. This reduction of pH makes both microorganisms and eggwhite proteins more heat stable. The stability of the highly heatlabile egg protein conalbumin is increased by the addition of salts of iron or aluminium, sufficiently to permit pasteurization at 60–62°C. Unlike their resistance to heat in other materials, bacteria are more resistant to heat in liquid whole egg at pH values of 5.6–5.6 than at values of 8.0–8.5. This is true when pH is lowered with an acid such as HCl. When organic acids such as acetic or lactic acid are used to lower pH, a decrease in heat resistance occurs.

## Proteins and other Substances

It is well known that proteins in the heating menstrum have a protective effect upon microorganisms. In order to achieve the same end results consequently high-protein-content foods must be heat processed to a greater degree than low-proteincontent foods in order to achieve the same end results. For identical numbers of organisms, the presence of colloidal-size particles in the heating menstrum also offers protection against heat. For example, under identical conditions of pH, numbers of organisms, and so on, it would take longer to sterilize pea puree than nutrient broth.

**Numbers of Organisms**

As the degree of heat resistance is higher, the larger is the number of organisms. It has been assumed that the mechanism of heat protection by large microbial populations was due to the production of protective substances excreted by the cells, and some authors claim to have demonstrated the existence of such substances. Many of the extracellular compounds in a culture media would be expected to be protein in nature, since proteins are known to offer some protection against heat and so they are consequently capable as affording some protection. Of perhaps equal importance in the higher heat resistance of large cell populations over smaller ones is the greater chance for the presence of organisms with differing degrees of natural heat resistance.

**Table 4.12. Effect of the number of spores of *Clostridium botulinum* on the thermal death time at 100°C.**

| *Number of Spores* | *Themal Death Time (Minutes)* |
|---:|:---:|
| 72,00,00,00,000 | 240 |
| 1,64,00,00,000 | 125 |
| 3,20,00,000 | 110 |
| 6,50,000 | 85 |
| 16,400 | 50 |
| 328 | 40 |

During the logarithmic phase bacterial cells tend to be less resistant white in the stationary phase of growth it tend to be most resistant. This is true for *S. senftenberg* whose stationary phase cells may be several times more resistant than log phase cells. Heat resistance has been reported to be high also at the beginning of the lag phase but decreases to a minimum as the cells enter the log phase. Old bacterial spores have been reported to be more heat resistant than young spores. At this time the mechanism of increased heat resistance of less active microbial cell is undoubtedly complex and is not well understood.

**Growth Temperature**

The heat resistance of microorganisms tends to increase as the temperature of incubation increases. Lechowich and Ordal showed that as sporulation temperature was increased for *B. subtilis* and *B. coagulans*, the thermal resistance of spores of both organisms also increased. It is conceivable that genetic selection favours the growth of more heat resistant strains at succeedingly high temperatures, although the precise mechanism of this effect is unclear. *S. senftenberg* grown at 44°C has been found to be approximately three times more resistant than cultures grown at 35°C.

**Inhibitory Compounds**

When heating takes place in the presence of heat resistant antibiotics, $SO_2$, and other microbial inhibitors as might be expected, a decrease in heat resistance of most microorganisms occurs. The use of heat+antibiotics and heat+nitrite together has been found to be more effective in controlling the spoilage of certain foods than either alone. The amount of heat that would be necessary if used alone is reduced by the practical effect of adding inhibitors to foods prior to heat treatment.

**Time and Temperature**

One would expect that the longer the time of heating, the greater the killing effect of heat. All too often, though, there are exceptions to this basic rule. A more dependable rule is that the higher the temperature, the greater the killing effect of heat. Time necessary to achieve the same effect decreases with the increase in temperature.

**Table 4.13. Effect of temperature upon the thermal death time of spores.**

| *Temperature* | *Clostridium botulinum (60,000,000,000 spores suspended in buffer at pH7)* | *A thermophile (150,000 spres per ml.of corn juice at pH 6.1) Minutes* |
|---|---|---|
| 100°C | 260 | 1140 |
| 105°C | 120 | |
| 110°C | 36 | 180 |
| 115°C | 12 | 60 |
| 120°C | 5 | 17 |

These rules assume that heating effects are immediate and not mechanically obstructed or hindered. Also important is the size of the heating vessel or container and its composition (glass, metal, plastic). Rather than smaller containers it should be obvious that it would take longer to effect pasteurization or sterilization in large ones. The same would be true of containers with walls that do not conduct heat as readilys as others.

## Relative Heat Resistance of Microorganisms

With the optimum growth temperatures the heat resistance of microorganisms is related in general. Psychrophilic microorganisms are the most heat sensitive of the three temperatures groups, followed by mesophiles and thermophiles. Sporeforming bacteria are more heat resistant than nonsporeformers, while thermophilic sporeformers are in general more heat resistant than mesophilic sporeformers. With respect to gram reaction, gram-positive bacteria tend to be more heat resistant than nonsporeforming rods. Yeasts ascospores is only slightly more resistant than vegetative yeasts though yeasts and molds tend to be fairly sensitive to heat. The asexual spores of molds tend to be slightly more heat resistant than mold mycelia.

Sclerotia are the most heat resistant of these types and sometimes survive and cause trouble in canned fruits. In the thermal processing of foods the heat resistance of bacterial endospores is of special interest. These structures are produced by *Bacillus* and *Clostridium* spp. usually upon the exhaustion of nutrients essential for continued vegetative growth, although other factors appear to be involved. Only one spore is produced per cell, and it may occur in various parts of the vegetative cell and possess various shapes and sizes, all of which are of taxonomic value. The endospore is not only resistant to heat but to drying, cold, chemicals, and other adverse environmental factors. Staining by ordinary methods is resisted by highly refractive body.

The refractivity is due in part to the spore coats, which consist of at least two layer outer (exine and inner (intine). Heat resistance is due also in part to the dehydrated nature of the cortex and spore core. Endospores are known to contain DNA, RNA, water, various enzymes, metal ions, and other compounds, especially dipicolinic acid (DPA). Numerous authors have related heat resistance to spore DNA and calcium content the precise mechanism of heat resistance of endospores is not yet fully understood. DPA may constitute 5-15 of the dry weight of endospores, while these bodies contain two to ten times more calcium than the corresponding vegetative cell. There is a general increase in heat resistance as the ratio of cation to DPA increases. The addition of chelating agents with high affinities toward calcium and manganese has been shown to decrease the heat resistance of endospores, while the addition of calcium and manganese generally restores thermal resistance. DPA, divalent cations, and ninbydrin-positive material into the medium release is accompanied by the thermal death of endospores.

**Thermal Destruction of Microorganisms**

It is necessary to understand certain basic concepts associated with this technology in order to better understand the thermal destruction of microorganisms relative to food preservation and canning. Below

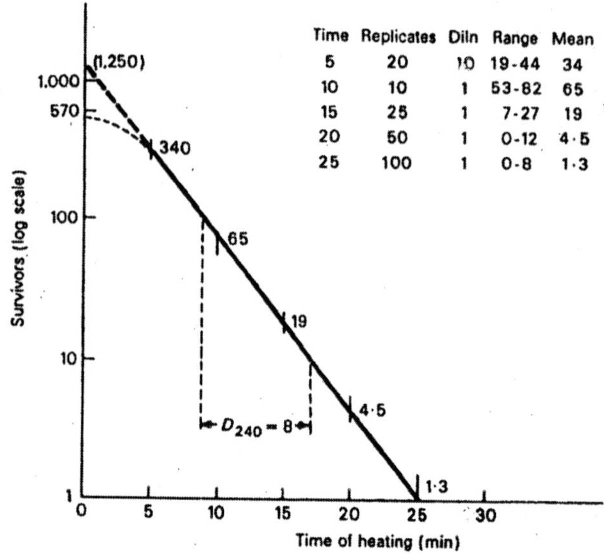


*Fig. 4.4. Rate of destruction curve. Spores of strain FS.7 heated at 240°F in canned pea brine pH 6.2.*

are listed some of the more important concepts, but for a more extensive treatment of thermobacteriology, the excellent monograph by Stumbo should be consulted.

## Thermal Dath Time (TDT)

To kill a given number of organisms at a specified temperature this is the necessary time. The time necessary to kill all cells is determined by this method by keeping the temperature constant. Of less importance is the thermal death point, which is the temperature necessary to kill a given number of microorganisms in a fixed time, usually 10 min. Various means have been proposed for determining TDT the tube, can, "tank," flask, thermo-resistometer, unsealed tube, and capillary tube methods.

In order to get the desired number of survivors for each test period a known number of cells is placed is a sufficient number of sealed containers which is the general procedure for determining TDT by these methods. The organisms are then placed in an oil bath and heated for the required time period. At the end of the heating period, containers are removed and cooled quickly in cold water. The organisms are then placed on a suitable growth medium, or the entire heated containers are incubated if the organisms are suspended in a suitable growth substrate. The specific organisms rows suitable when the suspensions or containers are incubated at a temperature. To form a visible colony death is defined as the inability of the organisms.

**D. Value**

This is the decimal reduction time, or the tie required to destroy 90% of the organisms. This value is numerically equal to the number of min required for the survivor curve to traverse one log cycle. It is a measure of the death rate of an organisms and mathematically, it is equal to the reciprocal of the slope of the survivor curve. When D is determined at 250°C, it is often expressed as D. D values of 0.20 to 2.20 min at 150°F have been reported for *S. aureus* strains, D 150°F of 0.50–0.60 min for *Coxiella burnetti*, and D 150°F of 020-0.30 for *Mycobacterium hominis*. D 95°C of 13.7 min has been found for *B. coagulans* while a D 95°C of 5.1 min has been reported for pH-elevating strains of *Bacillus licheniformis* spores in tomatoes.

**Table 4.14. Effect of pH on D values for spores of *C. botulinum* 62A suspended in three food prodcuts at 240°F.**

| | *D Value (in min)* | | |
|---|---|---|---|
| *pH* | *Spaghetti, tomato sauce, and cheese* | *Macaroni creole* | *Spanish rice* |
| 4.0 | 0.128 | 0.127 | 0.117 |
| 4.2 | 0.143 | 0.148 | 0.124 |
| 4.6 | 0.223 | 0.223 | 0.210 |
| 4.8 | 0.226 | 0.261 | 0.256 |
| 5.0 | 0.260 | 0.306 | 0.266 |
| 6.0 | 0.491 | 0.535 | 0.469 |
| 7.0 | 0.515 | 0.568 | 0.550 |

**Z Value**

To traverse one log cycle for thermal destruction curve required by degree by degree Fahemeheit is required to by the z value. Mathematically, this value is equal to the reciprocal of the slope of the TDT curve. While D reflects the resistance of an organisms to a specific temperature, z provides information on the relative resistance of an organism to different destructive temperature; it allows for the calculation of equivalent thermal processes at different temperatures. If for example, 3.5 min at 140°F is considered to be an adequate process and z =8.0, either 0.35 min at 148°F of 35 min at 132°F would be considered equivalent processes.

**F Value**

To destroy spores or vegetative cells of a particular organism this value with respect to its capacity is the equivalent time, in minutes at 250°F, of all heat considered. The integrated lethal value of heat received by all points in a container during processing is designated $F_s$ or $F_0$. This represents a measure of the capacity of a heat process to reduce the number of spores or vegetative cells of a given organism per container. When we assume instant heating and cooling throughout the container of spores, vegetative cells, or food, $F_0$ may be derived as follows: were a=number of cells in the initial population and b=number of cells in the final population.

**Thermal Death Time Curve**

Data are employed from Gillespy on the killing of flat sour spores at 240°F in canned pea brine at pH 62., for the purpose of illustrating a thermal destruction curve and D value. Counts were determined at intervals of 5 min with the mean viable numbers indicated below:

| *Time (min.)* | *Mean viable count* |
|---|---|
| 5 | 340.0 |
| 10 | 65.0 |
| 15 | 19.0 |
| 20 | 4.5 |
| 25 | 1.3 |

The number of survivors is plotted along the log scale to produce the TDT curve while on semi-log paper along the linear axis is plotted the time of heating in minutes which is presented. The curve is essentially linear, indicating that the destruction of bacteria by heat is logarithmic and obeys a first order reaction. Process calculation in the canning industry are based upon a logarithmic order of death though at either end of the TDT curve difficulty is encountered at times.

The relative resistance of spores or vegetative cells to heat is reflected by the use of D values. The most heat-resistant strains of *C. botulinum* types A and B spores have a D, value of 021, while the most heat-resistant thermophilic spores have D, values of around 4.0-5.0. Putrefactive anaerobe (PA.) 3679 was found by Stumbo et al. to have a D, value of 2.47 in cream-style corn, while flat sour (F.S.) spores 617 were found to have a D, of 0.84 in whole milk.

By use of D, values as below the approximate heat resistance of spores of thermophilic and mesophilic spoilage organisms may be compared.

| | |
|---|---|
| *B. stearothermophilus*: | 4.0–5.0 |
| *C. thermosaccharolyticum*: | 3.0–4.0 |
| *C. nigrificans*: | 2.0–3.0 |
| *C. botulinum* (types A and B): | 0.10–0.2 |
| *C. sporogenes* (including PA. 3679): | 0.10–1.5 |
| *B. coagulans:* | 0.01–0.07 |

The effect of pH and suspending menstrum on D values of *C. botulinum* spores are presented in Table. As noted above, microorganisms are more resitant at and around neutrality and show different derees of heat resistance in different foods. Along the linear axis are plotted the F degress while on the log scales are plotted the D values in order to determine the z value. From the data presented, the z value is seen to be 17.5 Values of z for *C. botulinum* range from 14.7 to 16.3, while for P.A. 3679 the range of 16.6-20.5 has bee reported. Some spores have been reported to have z values as high as 22. Peroxidase has been reported to have a z value of 47, while 50 has been reported for riboflavin and 56 for thiamine.

### 12-D Concept

The 12-D concept refers to the process lethality requirement long in effect in the canning industry and implies that the minimum heat process should reduce the probability of survival of the most resistant *C. botulinum* spores to $10^{-12}$. This concept is observed only for foods above pH 4.6 value since *C. botulinum* spores do not germinate and produce toxin below pH. 4.6. An example from Stumbo illustrates this concept from the standpoint of canning technology. If it is assumed that each container of food contains only one spore of *C. botulinuin*. Fo may be calculated by use of the general survivor curve equation with the other assumptions noted above in mind:

$$F_0 = D, (\log a - \log b)$$

$$F_0 = 0.21 (\log 1 - \log 10^{-12})$$

$$F_0 = 0.21 \times 12 = 2.52$$

Processing for 2.52 min at 250°F, then, should reduce the *C. botulinum* spores to one spore in one of one million containers. The potential number of *C. botulinum* spores is reduced even more, when it is considered that some flat-sour spores have D, values of about 4.0 and some canned foods recieve, $F_0$ treatment of 6.0–8.0.

### Aspetic Packaging

Followed by container closure and sterilization, nonsterile food is placed in nonsterile metal or glass containers in traditional canning methods. In aseptic packaging, sterile food under aseptic conditions is placed in sterile containers, and the packages are sealed under aseptic conditions as well. While the methodology of aseptic packaging was patented in the early 1960s, the technology was little until 1981 when the food and Drug Administration approved the use of hydrogen peroxide for the sterilization of flexible multilayered packaging materials used in aseptic processing systems. Asptically packaged are those foods in general that can be pumped through a heat exchange.

The widest application has been to liquids such as fruits juices, and a wide variety of singleserve products of this type has resulted. The technology for foods that contain particulates has been more difficult to develop, with microbiological considerations only one of the many problems to overcome. In determining the sterilization process for foods pumped through heat exchanges, the fastestmoving components (those with the minimum holding time) are used, and where liquids and particulates are mixed, the latter will be the slower moving. Making it more difficult to establish minimum process requirements that will effectively destroy both organisms and food enzymes as heat penetration rates are not similar for liquids and solids.

Some of the advantage of aseptic packaging are these: (1) products such as fruit juices are more flavourful and lack the metallic taste of those processed in metal containers; (2) flexible multilayered cartons can be used instead of glass or metal containers; (3) the time a product is subjected to high temperatures is minimized when ultrahigh temperatures are used; (4) the technology allows the use of membrane filtration of certain liquids; and (5) various container headspace gases such as nitrogen may be used. The output is lower than that for solid containers or the packages may not equivalent to glass or metal containers in preventing the permeation of oxygen, are among some of the disadvantages.

A wide variety of aseptic packaging techniques now exists, with more under development. Sterilization of packages is achieved in various ways, one of which involves the continuous feeding of rolls of packaging material into a machine where hot hydrogen peroxide is used to effect sterilization, followed by the forming, filling with food, and sealing of the containers. By a positive pressure of air or gas such as nitrogen sterlity of the filling operation may be maintained. Aseptically packaged fruit juices are shelf stable at ambient temperatures for 6 to 12 months or longer.

Food's spoilage in metal containers may be expected to differ from aseptically packaged foods. While hydrogen swells occur in high-

foods in the latter containers, aseptic packaging materials are nonmetallic. Other types of spoilage in low acids foods may be allowed by the permeation of oxygen by the nonmetal and nonglass containers but in aseptically packaged foods seam leakage may be expected to he absent.

## BY DRYING UP THE FOOD

In order to be active water is needed by the microorganisms and enzymes, this fact is considered while preserving foods by drying. In preserving foods by this method, one seeks to lower the moisture content of foods to a point where the activities of food spoilage and food-poisoning microorganisms are inhibited. Dried, desiccated, or low moisture (LM) foods are those that generally do not contain more than 25% moisture and have an $a_w$ between 0.00 and 0.60. These are the traditional dried foods. Freeze-dried foods are also in this category. Containing ranging between 15 and 50% moisture and an a,, between 0.60 and 0.85 is another category of shelf-stable foods. These are the intermediate moisture (IM) foods.

### Preparation and Drying of Low-Moisture Foods

Until drying have been achieved simply the fresh foods were exposed to sunlight in the earliest uses of food desiccation. Through this method of drying, which is referred to as sun drying, certain foods may be successfully preserved if the temperature and relative humidity (R.H.) allow. Fruits such as grapes, prunes, figs, and apricots may be dried by this method, which requires a large amount of space for large quantities of the product.

The drying methods of greatest commercial importance consist of spray, drum, evaporation, and freeze drying. With a few exceptions the same manner as that of freezing are adopted for handling foods in preparatory to drying. In the drying of fruits such as prunes, alkali dipping is employed by immersing the fruits into hot lye solutions of between 0.1 and 1.5%. This is especially true when sun drying is employed. In order that level of between 1,000 and 3,000 ppm may be absorbed, light-coloured fruits and certain vegetables are treated with $SO_2$. The latter treatment helps to maintain colour, conserve certain vitamins, prevent storage changes, and reduce the microbial load.

After drying, fruits are usually heat pasteurized at 150-185°F for 30-70 min. Similar to the freezing preparation of vegetable foods, blanching or scalding is a vital step prior to dehydration. This may be achieved by immersion from 1 to 8 min, depending upon the particular type of product. The primary function of this step in drying is to destroy enzymes that may become active and bring about undesirable changes in the finished products. Leafy vegetables generally require less time than peas, beans, or carrots. For drying, temperatures of 140–145°F have been found to be safe for many vegetables. The moisture content of vegetables should be reduced below 4% in order to have satisfactory storage life and quality. Many vegetables may be made more stable if given a treatment with $SO_2$ or a sulfite.

The drying of vegetables is usually achieved by used to tunnel, belt, or cabinet-type driers. Before being dehydrated meat is usually cooked. After drying the final moisture content should be approximately 4% for beef and pork. Milk is dried as either whole milk or nonfat skim milk. The dehydration may be accomplished by either the drum or spray method. The removal of about 60% water from whole milk results in the production of evaporated milk, which has about 11.5% lactose in solution. Sweetened condensed milk is produced by the addition of sucrose or glucose before evaporation so that the total average content of all sugar is about 54%, or over 64% in solution.

The fact that the sugars tie up some of the water and make it unavailable for microbial growth is revealed by the stability of sweetened condensed milk is due in part. Eggs may be dried as whole egg powder, yolks, or egg white. Prior to drying by reducing the glucose content dehydration stability is increased and the most commonly employed method is spray drying.

In freeze drying (lyophilization, cryophilization) actual freezing is preceded by the blanching of vegetables and the precooking of meats. The rate at which a food material freezes or thaws is influenced by the following factor (1) the temperature differential between the product and the cooling or heating medium. (20 the means of transferring heat energy to, from, and within the product (conduction, convection, radiation), (3) the type, size, and shape of the package, and (4) the size, shape, and thermal properties of the product. Rapid freezing has been shown to produce products that are more acceptable than slow freezing. As discussed in Chapter rapid freezing allows for the formation of small ice crystals and consequently less mechanical damage to food structure. Characteristics more like the fresh product than slow frozen foods in general in displayed by fastfrozen foods which take up more water on thawing. By sublimation the water in the form of ice is removed after freezing. This process is achieved by various means of heating plus vacuum. Freezable and unfreezable are the two groups into which the water content of protein foods can be placed. Unfreezable (bound) water has been defined as that which remains unfrozen below-30°C. The removal of freezable water takes place during the first phases of drying and this phase of drying may account for the removal of anywhere from 40 to 95% of the total moisture. Generally bound water is the last water to be removed, as which some may be removed, throughout the drying process. Unless heat treatment is given prior to freeze drying, freeze-dried foods retain their enzymes. 40–80% of the enzyme activity is not destroyed and may be retained after 16mon storage at–20°C. The final product moisture level in freezedried foods may be about–8%, or have an $a_w$ of 0.10–025.

For high temperature vaccum-drying, freeze drying is generally preferred. Among the disadvantages of the latter compared to the former are the following:

1. Pronounced shrinkage of solids.
2. Migration of dissolved constituents to the surface when drying solids.
3. Extensive denaturation of proteins.
4. Case-hardening: the formation of a relatively hard, impervious layer at the surface of a solid is caused by one or more of the first three changes; this impervious layer slows rates of both dehydration and reconstitution.
5. Formation of hard, impervious solids when drying liquid solution.
6. Undesirable chemical reactions in heat-sensitive materials.
7. Excessive loss of desirable volatile constituents.
8. Difficulty of rehydration as a result of one or more of the above changes.

### Effect of Drying Upon Microorganisms

Although some microorganisms are destroyed in the process of drying, this process is not lethal per see to microorganisms, and indeed, many types may be recovered from dried foods, especially if poor-quality foods are used for drying and if proper practices are not followed in the drying steps. Yeasts require less and mold still lesser level of moisture for their growth while bacteria require relatively high level which is previously indicated. Since most bacteria require $a_w$ values above 0.90 for growth, these organisms play no role in the spoilage of dried foods.

With respect to the stability of dried foods, Scott has related a, levels to the probability of spoilage in the following manner. At $a_w$, values of between 0.80 and 0.85, spoilage occurs readily by a variety of fungi in from 1 to 2 weeks. At $a_w$ values of 0.75, spoilage is delayed, with fewer types of organisms in those products that spoil.

During prolonged holding spoilage may to occur and at a. 0.70 it is greatly delayed. At $a_w$ of 0.65, very few organisms are known to grow, and spoilage is most unlikely to occur for even up to 2 years. Some authors have suggested that dried foods to be held for several years should be processed so that the final $a_w$ is between 0.65 and 0.75, with 0.70 suggested be most. Feasts and molds are the organisms which are most likely to grow at $a_w$ levels of about 0.90. This value is near the minimum for most normal yeasts. Even though spoilage is all but prevented at $a_w$<0.65, some molds are known to grow very slowly at $a_w$ 0.62–0.62.

**Table 4.15. Minimum $a_w$ reported for the germination and growth of various food spoilage yeasts and molds.**

| *Organism* | *Minimum $a_w$* |
|---|---|
| Candida utilis | 0.94 |
| Botrytis cinerea | 0.93 |
| Rhizopus stolonifer (nigricans) | 0.93 |
| Nucor spinosus | 0.93 |
| Candida scottii | 0.92 |
| Trichosporon pullulnas | 0.91 |
| Candida zeylanoides | 0.90 |
| Sacharomyces rouxii | 0.90 |
| Endomyces vernalis | 0.89 |
| Alternaria citri | 0.84 |
| Aspergillus glaucus | 0.70 |
| Aspergillus echinulatus | 0.64 |
| Sacharomyces rouxii (Z.borkeri) | 0.62 |

Osmophilic yeasts such as *Saccharomyces rouxii* strains have been reported to grow at an $a_w$, of 0.65 under certain conditions. Molds are the most troublesome group of microorganisms in dried foods, with the *Aspergillus glaucus* group being the most notorious at low $a_w$ values. The minimum $a_w$ values reported for the germination and growth of molds and yeasts are presented in Table. Pitt and Christian found the predominant spoilage molds of dried and high-moisture prunes to be members of the *A. glaucus* group and *Xeromyces bisporus*. Aleuriospores of X. bisporus were able to germinate in 120 days at an $a_w$ of 0.095. For both asexual and sexual sporulation generally higher moisture levels were required.

The "alarm water" content has been suggested as a guide to the storage stability of dried foods. If molds growth is to be avoided, the alarm water content which is the water content should not be exceeded. While these values may be used to advantage, they should be followed with caution, as a rise of only 1% may be disastrous in some instances. The "alarm water" content for some miscellaneous foods is presented in Table. In freeze-dried foods, the rule of thùmb has been to reduced the moisture level to 2%. Bürker and Decareau have pointed out that this low level is probably too severe for some foods that might keep well at higher levels of moisture without the extra expense of removing the last low levels of water.

**Table 4.16. The "alarm water" content for miscellaneous foods, assuming R.H. of 70% and a temperature of 20°C.**

| *Foods* | *% water* |
|---|---|
| *Whole milk powder* | ca. *8* |
| Dehydrated whole eggs | 10–11 |
| Wheat flour | 13–15 |
| Rice | 13–15 |
| Milk powder (separated) | 15 |
| Fat free dehydrated meat | 15 |
| Pulses | 15 |
| Dehydrated vegetables | 14–20 |
| Starch | 18 |
| Dehydrated fruit | 18–25 |

As yeasts, molds and many gram-negative and positive bacteria survives so do bacterial endospores though some microorganisms are destroyed by drying. In their study of bacteria from chicken meat after freeze drying and rehydration at room temperature. May and Kelly were able to recover about 32% of the original flora. Under certain conditions *S. aureus* added prior to freeze drying could survive as showed by these workers. Some or all foodborne parasites, such as *Trichinella spiralis*, have been reported to survive the drying process. The present goal is to produce dried foods with a total count of not more than 100,000/g.

It is generally agreed that the coliform count of dried foods should be zero or nearly so, and no foodpoisoning organisms should be allowed, with the possible exception of low numbers of *C. perfringens* and *S. faecalis*. Relatively fewer organisms are destroyed during the freezedrying process with the exception of those that may be destroyed by blanching or precooking. More are destroyed during freezing than during dehydration. During freezing, between 5 and 10% of water remains "bound" to other constituents of the medium. This water is removed by drying. Death or injury from drying may result from: (1) denaturation in the still frozen, undried portions due to concentration resulting from freezing, (2) the act of removing the "bound" water, and/or (3) recrystallization of salts or hydrates formed from eutectic solutions. When death occurs during dehydration, the rate is highest during the early stages of drying. Foung cultures have been reported to be more sensitive to drying than old cultures. The freeze-drying method is to course, one of the best known ways of preserving microorganisms. The cells may remain viable indefinitely once the process has been completed. David found on examining the viability of 277 cultures of bacteria, yeasts and molds that had been lyophilized for 21 years that only 3 failed to survive.

**Storage Stability of Dried Foods**

Upon holding food becomes undesirable which is resulted due to certain chemical changes which are subject desiccated foods, in the absence of fungal growth. In dried foods that contain fats and oxygen, oxidative rancidity is a common form of chemical spoilage. Foods that contain reducing sugars undergo a coulour change known at the Maillard reaction or nonenzymic browning. This process is brought about when the carbonyl groups of reducing sugars react with amino groups of proteins and amino acids, followed by series of other more complicated reactions. Not only because of the bitter taste imparted to susceptible foods but also unnatural coulour, maillard type browning is quite undesirable in fruits and vegetables. If the moisture content is above 2% browning is undergone by freezed-dried foods. Thus, the moisture content should he held below 2%. With regard to $a_w$, the maximal browning reaction rates in fruits and vegetable products occur in the 0.6–0.75 range, while for nonfat dry milk browning, it seems to occur most readily at about 0.70.

A loss of vitamin C in vegetables, general discolourations, structural changes leading to the inability of the dried product to fully rehydrate, and toughness in the rehydrated, cooked product are other chemical changes that takes place in dried foods. So that preventative measures against one are also effective against others to varying degrees conditions that favour one or more of the above chances is dried foods generally tend to favour all. At least four methods of minimizing chemical moisture content as low as possible. Gooding has pointed out that lowering the moisture content of cabbage from to 3% doubles its storage life at 37°C. (2) Reduce the level of reducing sugars as low as possible. These compounds, of course, are directly involved in nonenzymic browning and their reduction has been shown to increase storage stability. (3) When blanching, use water in which the level of leached soluble solids is kept low. Gooding has shown that the serial blanching of vegetables in the same water increases the chances of browning. The explanation given is that the various extracted solutes (presumably reducing sugars and amino acids) are impregnated on the surface of the treated products at relatively high levels. (4) Use of sulfur dioxide.

The treatment of vegetables prior to dehydration with this gas protects vitamin C along with retarding the browning reaction. It is not well understood what is the precise mechanism of this gas in retarding the browning reaction, but it apparently does not block reducing groups of hexoses. It has been suggested that it may act as a free radical acceptor. R.H. of the storage environment is one of the most important considerations in preventing fungal spoilage of dried foods. If improperly packed and stored under conditions of high R.H., dried foods will pick up moisture from the atmosphere until some degree of equilibrium has been established. Since the first part dried product to gain moisture is the surface, spoilage would be inevitable, as surface growth tends to be characteristic of mods due to their oxygen requirements.

**Intermediate-moisture Foods (IMF)**

With moisture content of around 15 to 50% and an $a_w$ between 0.60 and 0.85 *intermediate-moisture foods* (IMF) are characterized, which is noted earlier in this chapter. These foods are shelf stable at ambient temperatures for varying periods of time. While impetus was given to this class of foods during the early 1960s with the development and marketing of intermediate-moisture dog food, foods for human consumption that meet the basic criteria of this class have been produced for many years. To distinguish them from the newer IMFs these are referred to as "traditional IMFs". In Table 4.17 are listed some traditional IMFs along with their, $a_w$ values. All of these foods have, of course, lowered $a_w$ values, which are achieved by withdrawal of water by desorption, adsorption, and/or the addition of permissible additives such as salts and sugars. Not only by $a_w$ values of 0.60-0.85 but by the use of additives such as glycerol, glycols, sorbitol, sucrose, and so forth, as humectants, and by their content of fungistats such as sorbate and benzoate is characterized the newly developed IMFs. The remainder of this chapter is devoted to the newly developed IMFs.

**Table 4.17. Some traditional intermediate moisture foods ($a_w$, values taken from the literature).**

| *Food products* | *$a_w$ range* |
|---|---|
| Dried fruits | 0.60–0.75 |
| Cake and pastry | 0.60–0.90 |
| Forzen foods | 0.60–90 |
| Sugars, syrups | 0.60–75 |
| Some candies | 0.60–0.65 |
| Commercial pastry fillings | 0.60–71 |
| Cereals (some) | 0.65–0.75 |
| Fruit cake | 0.73–0.83 |
| Honey | 0.75 |
| Fruit juice concentrates | 0.79–0.84 |
| Fams | 0.80–0.91 |
| Sweetened condensed milk | 0.83 |
| Fermented sausages (some) | 0.83–0.87 |
| Maple syrup | 0.90 |
| Ripened Cheese (some) | 0.96 |
| Liverwurst | 0.96 |

## Preparation of IMF

At $a_w$ values 5 near 0.86 grows the only bacterium S. aureus of public health importance, IMF can be prepared by: (1) formulating the product so that its moisture content is between 15 and 50 percent, (2) adjusting the a,, to a value below 0.86 by use of humectants, and (3) adding an antifungal agent to inhibit the rather large number of yeasts and molds that are known to be capable of growth at $a_w$ values above 0.70. Additional storage stability is achieved by reducing pH. The actual process and the achievement of storage stability of the product are considerably more complicated though this is essentially all that one needs to produce an IMF.

The determination of the $a_w$, of a food system is discussed. One can use also Raoult's law of mole fractions where the number of moles of water a solution is divided by the total number of moles in the solution:

$$a_w = \frac{\text{Moles of } H_2O}{\text{Moles of } H_2O + \text{Moles of solute}}$$

For example, a liter of water contains 55.5 moles. Assuming that the water is pure.

$$a_W = \frac{55.5}{55.5+0} = 10.0$$

If however, one mole of sucrose is added,

$$a_W = \frac{55.5}{55.5+1} = 0.98$$

To give a specified $a_w$ value this equation can be rearrenged to solve for the number of moles of solute required. While the foregoing is not incorrect, it is highly oversimplified, since food systems are complex by virtue of their content of ingredients that interact with water and with each other in ways that are difficult to predict. Sucrose, for example, decreases $a_w$ more than expected by the above so that calculations based upon Raoult's law may be meaningless. The concern of several investigators (3, 16, 31) in the development of techniques and methods to predict $a_w$ in IMF more accurately, while an extensive evaluation of available $a_w$ measuring instruments and techniques has been carried out by 1 abuza et al. Either by adsorption or desorption water may be removed in preparing MO.

By adsorption, food is first dried (often freeze dried), and then subjected controlled rehumidification until the desired composition is achieved. By desorption, the food is placed in a solution of higher osmotic pressure so that at equilibrium, the desired $a_w$ is reached. While identical $a_w$ values may be achieved by these two methods. IMF produced by adsorption is more inhibitory to microorganisms than that produced by desorption. When sorption isotherms of food materials are determined, adsorption isotherms sometimes reveal that less water is held than for desorption isotherms at the same $a_w$ The sorption isotherm of a food material is a plot of the amount of water adsorbed as a function of the relative humidity or activity of the vapour space surrounding the material. After equilibrium has been reached at a constant temperature it is the amount of water that is held.

Sorption isotherms may be either adsorption or desorption, and when the former procedure results in the holding of more water that the latter, the difference is ascribed to an hysteresis effect. Not to be dealt further here and already been discussed by labuza sloan et al. and others is the hysteresis effect as well as other physical properties associated with the preparation of IMF. The overall IMF preparation procedures all add to the complication by the sorption properties of an

IMF recipe, the ingredients and the order of mixing of ingredients and may result in both direct and indirect effects on the microbiology of these products.

The general techniques employed to change the water activity in producing an IMF are summarized below:

1. Moist infusion. Solid food pieces are soaked and/or cooked in an appropriate solution to give the final product the desired water level (desorption).
2. Dry infusion. Solid food pieces are first dehydrated, following which they are infused by soaking in a solution containing the desired osmotic agents (adsorption).
3. Component blending. All IMF compounts are weighed, blended, cooked, and extruded or otherwise combined to give the finished product the desired $a_w$.

**Table 4.18. Typical composition of soft moist or intermediate moisture dog food.**

| *Ingredient* | % |
|---|---|
| Meat by-products | 32.0 |
| Soy flakes | 33.0 |
| Sugar | 22.0 |
| Skimmed milk dry | 2.5 |
| Calcium and phosphorus | 3.3 |
| Propylene glycol | 2.0 |
| Sorbitol | 2.0 |
| Animal fat | 1.0 |
| Emulsifier | 1.0 |
| Salt | 0.6 |
| Potassium sorbate | 0.3 |
| Minerals, vitamins and colour | 0.3 |
| | 100.0% |

The 1-cm-thick slices equilibrated following cooking at 95°–100°C in water and holding overnight in a refrigerater. Equilibration is possible without cooking over prolonged periods under refrigeration IMF deep-fried catfish, with raw samples of about 2 g each, has been prepared by the moist infusion method. Pet foods are more often prepared by component blending. The general composition of one such product is given in Table. The general way in which a product of this type is made is as follows. With liquid ingredients we mixed and grounded the meat and meat products and lately with dry ingredient mix (salts, sugars, dry solids, and so on) is cooked or heat treated and later mixed the resulting slurry. Once the latter are mixed into the slurry, an additional cook or heat process may be applied prior to extrusion and packaging. The extruded material may be further shaped in the form of patties or packaged in loose form.

According to Acott and Labuza this is an adaptation of pemmican, an Indian trail and winter storage food made of buffalo meat and berries. To the chickenbased IMF the name given is Hennican. By adjustment of ingredient mix both moisture content and a,, of this system can be altered. The humectants commonly used in pet food formulations are propylene glycol, polyhydric alcohols (sorbitol, for example), polyethylene glycols, glycerol, sugars (NaCl, KCl, and so on). Propylene glycol, K-sorbate, Na-benzoate, and others are he commonly used mycostats. The pH of these products may be as low as 5.4 and as high as 7.0.

**Table 4.19. Composition of IM food hennican.**

| *Components* | *Amount (wt basis, %)* |
|---|---|
| Raisins | 30 |
| Water | 23 |
| Peanuts | 15 |
| Chicken (freeze dried) | 15 |
| Non-fat dry milk | 11 |
| Peanut butter | 4 |
| Honey | 2 |

## Microbial Aspects of IMF

Whether that gram-negative bacteria will proliferate is made unlikely by the general a. range of IMF products. This is true also for most gram-positive bacteria with the exception of cocci, some sporeformers, and lactobacilli. In addition to the inhibitory effect of lowered $a_w$. antimicrobial activity results from an interaction of pH, Eh, added preservatives (including some of the humectants), the competitive microflora, generally low storage temperatures, and the pasteurization or other heat processes applied druing processing.

In this desorption IM pork at $a_w$ 0.88, the numbers remained stationary for about 15 days and then increased slightly, while in the adsorption IM system at the same $a_w$. the cells died off slowly during the first three weeks and thereafter more rapidly. At all $a_w$ values below 0.88, the organisms died off, with the death rate considerably higher at 0.73 than at higher values. Findings similar to these have

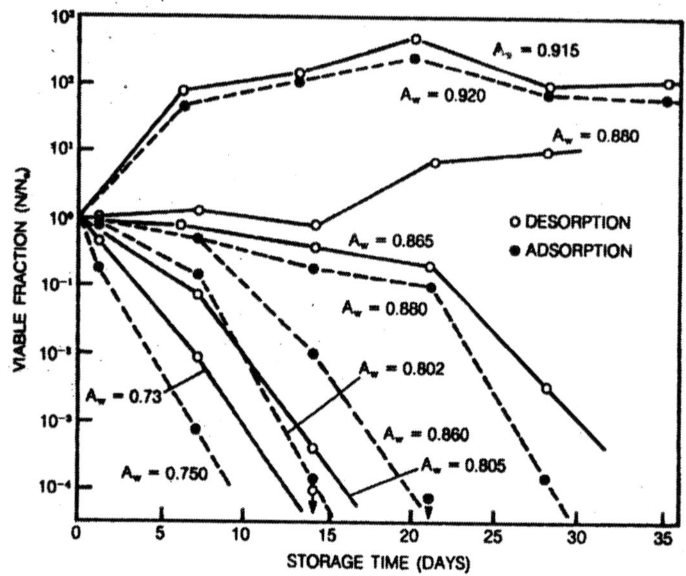


*Fig. 4.5. Vaibility of Staphylococcusaures in IMF system: pork cubes and glycerol at 25°C.*

been reported by Haas et al. who found that an inoculum of $10^5$ staphylococci in a meat-sugar system at $a_w$ 0.80 decreased to $3 \times 10^3$ after 6 days and to $3 \times 10^2$ after one month. Below $a_w$ 0.86 enterotoxin is not produced though at an $a_w$ of appears that enterotoxin A is produced at lower values of $a_w$ than enterotoxin B.

That the effectiveness of the IM system against S. aureus F 265 was a function of both pH and $a_w$ is showed by using the model IM Hennican at pH 5.6 and $a_w$ 0.91 Boylan et al. As noted above, adsorption systems are more destructive to microorganisms than desorption systems. When desorption system are involved reported minimum arts apply in IMF-system and if adsorption method is adspted in food preparation growth minima are much higher is found by Labuza *et al. S. aureus* was inhibited at $a_w$ 0.9 in adsorption while values between 0.75 and 0.84 were required for desorption systems. A similar effect was noted for molds, yeasts, and pseudomonads.

In regard to the effect of IMF systems on the heat destruction of bacteria, it is noted that heat resistance increases as $a_w$ is lowered and that the degree of resistance is dependent upon the compounds employed to control $a_w$. In a study of the death rate of salmonellae and staphylococci in the IM range of about 0.8 at pasteurization temperatures it has been found that cell death occurs under first order kinetics. Findings of many others have been confirmed by these investigators that the heat destruction of vegetative cells is at a minimum in the IM range especially when solid menstrum is employed.

A dry type product would result but if a, were reduced to around 0.70, these products would be made quite stable with respect to molds in IMF systems: A large number of molds are capable of growth in the 0.80 range, and the shelf life of IM pet foods is generally limited by the growth of these organisms. The interaction of various IM parameters on the inhibition of molds was shown by Acott et al. In their evaluation of seven chemical inhibitors used alone and in combination to inhibit Aspergillus niger and A. glaucus inocula, propylene glycol was the only approved agent that was effective alone. None of the agents tested could inhibit alone at $a_w$ 0.88, but in combination the product was made shelf stable. All inhibitors were found to be more effective at pH 5.4 and $a_w$ 0.85 than at pH 6.3. Growth of the two fungi occurred in 2 weeks in the $a_w$ 0.85 formulation without inhibitors but did not occur until 25 weeks when Ksorbate and Ca-propionate were added. With the inhibition being greater at $a_w$ 0.85 than at 0.88 both fungistats inhibited the growth of Staphylococcus. This is probably an example of the combined effect growth inhibition of microorganisms in IMF systems as previously noted.

**Storage stability of IMF**

The undesirable chemical changes that occur in dried foods occur also in IM foods. Lipid oxidation and Maillard browning are at their optima in the general IMF ranges of $a_w$ and percent moisture. A recent report, however, indicates that the maximum rate for Maillard browning occurs in the 0.4-0.5 $a_w$ range, especially when glycerol is used as the humectant.

In preventing moldiness and for overall shelf stability the storage of IMFs under the proper conditions of humidity is imperative. The measurement of equilibrium relative humidity (ERH) is of importance in this regard ERH is an expression of the desorbable water present in a food product, and is further defined by the following equation:

$$ERH = (P_{equ}/P_{sat})\ T,\ P = 1\ \text{atm}$$

where $P_{equ}$ is partial pressure of water vapour in equilibrium with the sample in air at 1 atmosphere total pressure and temperature T.P.. is the saturation partial vapour pressure of water in air at a total pressure of 1 atmosphere and temperature T. Until the equilibrium partial pressure at that temperature is equal to the partial pressure of water in the moist a food in moist air exchanges water so that the ERH value is a direct measure of whether moisture will be sorbed or desorbed. In the case of foods packaged or wrapped in moisture-impermeable materials, the relative humidity of the food-enclosed atmosphere is determined by the ERH of the product, which in turn in controlled by the nature of the dissolved solids present, ratio of solids to moisture, and the like. Both traditional and newer IMF products have longer shelf stability under conditions of lower ERH.

**Table 4.20. Time for growth of micrbes in inoculated dog food with inhibitors, pH 5.4.**

| | *Storage conditions* | |
|---|---|---|
| *Inhibitor* | *$a_w$ 0.85*<br>*9-mon storage* | *$a_w$=0.88*<br>*6-mon storage* |
| No inhibitor added | A. niger—2 *wk*<br>A. glaucus—1 wk<br>S. epider—2 wk | A. niger—1 *wk*<br>A. glaucus—1 *wk*<br>S. epider—1/2 *wk* |
| K-sorbate (0.3%) | *No mold*<br>S. epider—25 *wk* | A. niger—5 *wk*<br>*S.* epider-3½ *wk* |
| Ca-propionate (0.3%) | A. niger—25 *wk*<br>A. glaucus- 25 *wk*<br>S. epider—3½ *wk* | A. glaucus—2 *wk*<br>*S.* epider-1½ *wk* |

Gas impermeable packaging affects the Eh of packaged products in addition to the direct effect of packaging on ERH, with consequent inhibitory effects upon the growth of aerobic microorganisms.

## Preservation by Radiation

In their search for new, improved methods of food preservation, investigators have paid special attention to the possible utilization of radiations of various frequencies, ranging from low-frequency electrical current to high-frequency gamma rays. Much of this work has focused on the use of ultraviolet radiation, ionizing radiation, and microwave heating.

It is common to group the entire spectrum of radiation into two categories, one on each side of visible light. Low-frequency, long-wavelength, low-quantumenergy radiation ranges from radio waves to infrared. The effect of these radiations on microorganisms is related to their thermal agitation of the food. Conversely, the high-frequency, shorter-wavelength radiations have high quantum energies and actually excite or destroy organic compounds and microorganisms without heating the product. Microbial destruction without the generation of high temperatures suggested the term "cold sterilization."

When applied to the food industry, shorter-wavelength radiation can be further divided into two groups. Lower-frequency and lower-energy radiation, for example, the ultraviolet part of the spectrum, has sufficient energy only to excite molecules. This area of the spectrum is employed in the food industry and is covered in the section on ultraviolet irradiation. Radiations of higher frequencies have high energy contents and are capable of actually breaking individual molecules into ions, hence the term ionizing irradiation.

## Ultraviolet Irradiation

Of the various electromagnetic radiations, ultraviolet irradiation has been the most widely used in the food industry. Radiation with wavelengths near 260 nm is absorbed strongly by purines and pyrimidines and is therefore the most germicidal. Ultraviolet radiation around 200 nm is strongly absorbed by oxygen, may result in the production of ozone, and is ineffective against microorganisms.

### Germicidal Lamps

The usual source of ultraviolet radiation in the food industry is from quartzmercury vapor lamps or low-pressure mercury lamps, which emit radiation at 254 rim. Radiation from these lamps includes rays in the visible range and those in the erythemic range, which have an irritating effect on skin and mucous membranes. The lamps are available in various sizes, shapes, and power. The newer types release only negligible amounts of ozone.

### Factors Influencing Effectiveness

It should be emphasized that only direct rays are effective unless they comc from special reflectors, and even then their effectiveness is reduced. The factors that influence the effectiveness of ultraviolet rays are as follows:

#### *Time*

The longer the time of exposure to a given concentration, the more effective the treatment.

#### *Intensity*

The intensity of the rays reaching an object will depend on the power of the lamp, the distance from the lamp to the object, and the kind and amount of interfering material in the path of the rays. Obviously, the intensity will increase with the power of the lamp. Intensity is usually measured as microwatts per square centimeter ($\mu$ W/cm$^2$). The quantity or dose of irradiation actually absorbed by an organism or a product thus is expressed by the product of the time

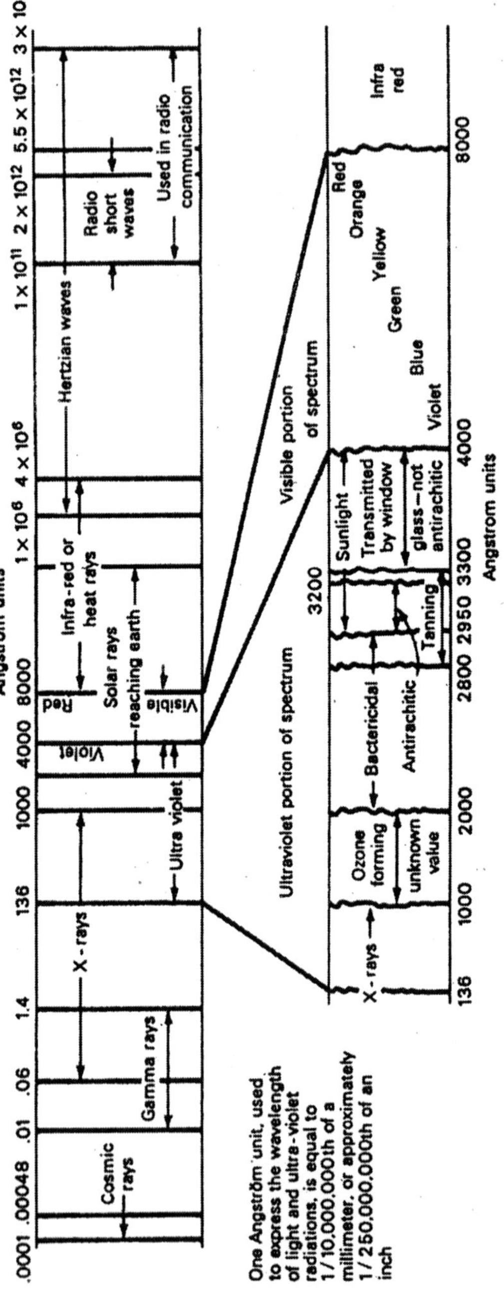


*Fig. 4.6. Spectrum charts.*

and intensity. Within the short distances common in industrial uses, the intensities of the rays vary inversely with distance from the lamp. A lamp is about 100 times as effective in killing microorganisms at 5 in, than at 8 ft from the irradiated object. Most tests are reported from a distance of about 12 in. Dust in the air or on the lamp reduces the effectiveness, as does too much atmospheric humidity. Over 80 percent relative humidity definitely reduces the penetration through air, but humidities below 60 percent have little effect.

***Penetration***

The nature of the object or material being irradiated has an important influence on the effectiveness of the process. Penetration is reduced even by clear water, which also exerts a protective effect on microorganisms. Dissolved mineral salts, especially of iron, and cloudiness greatly reduce the effectiveness of the rays. Even a thin layer of fatty or greasy material cuts off the rays. There is no penetration through opaque material. Therefore, the rays affect only the outer surface of most irradiated foods directly exposed to the lamp and do not penetrate to microorganisms inside the food. The lamps do serve, however, to reduce the number of viable organisms in the air surrounding the food.

**Effects on Humans and Animals**

Gazing at ultraviolet lamps produces irritation of the eyes within a few seconds, and longer exposure of the skin results in erythema, or reddening. The effect on animals is usually not as marked, although the eyes, especially of chicks, may be irritated.

**Action on Microorganisms**

As has been stated, the intensity of the rays when they reach the organism, the time in which they act, and the location of the organism determine the germicidal effect. Each microorganism has a characteristic resistance to ultraviolet irradiation. This can vary with the phase of growth and the physiological state of the cell. It takes as much as five times the exposure to kill vegetative cells of some bacteria compared with others, but in general the killing exposure does not vary widely among different species. The location of the organism during the tests has a marked influence. For example, 97 to 99 percent of *Escherichia coli* in air were killed in 10 sec at 24 in with a 15-watt lamp, but 20 sec at 11 in was necessary for bacteria on the surface of an agar plate. Capsulation or clumping of bacteria increases their resistance. Bacterial spores usually take from two to five times as much exposure as the corresponding vegetative cells. Some types of pigmentation also have a protective effect. Generally, yeasts are from two to five times as resistant as bacteria, although some are easily killed. The resistance of molds is reported to be from ten to fifty times that of bacteria. Pigmented molds are more resistant than nonpigmented molds, and spores are more resistant than mycelium. The killing effect of ultraviolet rays is usually explained by the "target theory," which is described in the discussion of ionizing radiation.

**Table 4.21. Ultraviolet radiation doses to destroy certain groups of microorganisms.**

| *Microorganism* | *Dose needed for 1 log cycle reduction or 1 D value ($\mu W$ sec $\times 10^3$)* |
|---|---|
| *Gram-negative bacteria* | |
| Facultative anaerobes | 0.8–6.4 |
| Aerobes | 3.0–5.5 |
| Phototrophic | 5.0–6.0 |
| *Gram-positive bacteria* | |
| *Bacillus* | 5.0–8.0 |
| *Bacillus* spores | 8.0–10.0 |
| Micrococcus | 6.0–20.0 |
| *Staphylococcus* | 2.2–5.0 |
| *Molds* | 10.0–200.0 |
| *Yeasts* | 3.0–10.0 |

## Applications in the Food Industry

The use of ultraviolet irradiation in the food industry will be discussed in connection with the preservation of specific foods. Examples of the successful use of these rays include treatment of water used for beverages; aging of meats; treatment of knives for slicing bread; treatment of bread and cakes; packaging of sliced bacon; sanitizing of eating utensils; prevention of growth of film yeast on pickle, vinegar, and sauerkraut vats; killing of spores on sugar crystals and in sirups; storage and packaging of cheese; prevention of mold growth on walls and shelves; and treatment of air used for, or in, storage and processing rooms.

# IONIZING RADIATIONS

## Kinds of Ionizing Radiations

Radiation classified as ionizing includes x-rays or gamma rays, cathode or beta rays, protons, neutrons, and alpha particles. Neutrons result in residual radioactivity in foods, and protons and alpha particles have little penetration. Therefore, these rays are not practical for use in food preservation and will not be discussed.

X-rays are penetrating electromagnetic waves which are produced by bombardment of a heavy-metal target with cathode rays within an evacuated tube. They are not currently considered economical for use in the food industry.

*Gamma rays* are like x-rays but are emitted from by-products of atomic fission or from imitations of such by-products. Cobalt 60 and cesium 137 have been used as sources of these rays in most experimental work thus far, with cobalt 60 being the most promising for commercial applications.

*Beta rays* are streams of electrons (beta particles) emitted from radioactive material. *Electrons* are small, negatively charged particles of uniform mass that form part of the atom. They are deflected by magnetic and electric fields. Their penetration depends on the speed with which they hit the target. The higher the charge of the electron, the deeper its penetration.

*Cathode rays* are streams of electrons (beta particles) from the cathode of an evacuated tube. In practice, these electrons are accelerated by artificial means.

**Definition of Terms**

Before the utilization of ionizing radiations can be discussed, a few terms must be defined.

A *roentgen* (*r*) is the quantity of gamma or x-radiation which produces one electrostatic unit of electric charge of either sign in a cubic centimeter of air under standard conditions.

A *roentgen-equivalent-physical* (*rep*) is the quantity of ionizing energy which produces, per gram of tissue, an amount of ionization equivalent to a roentgen. A *megarep* is 1 million *rep*. One *r*, or 1 *rep*, is equivalent to the absorption of 83 to 90 erg per gram of tissue.

The *rad* now is employed chiefly as the unit of radiation dosage, being equivalent to the absorption of 100 erg per gram of irradiated material. A *megarad* (Mrad) is 1 million rad, and a *kilorad* (Krad) is 1,000 rad.

An *electronvolt* (eV) is the energy gained by an electron in moving through a potential difference of I volt. A *meV* is 1 million electronvolts.

A meV, then, is a measure of the intensity of the irradiation, and a rep is a measure of the absorbed energy that is effective within the food.

A *Gray* (Gy) equals 100 rads and is being used as a term to replace rads in some references.

*Radappertization* is a term used to define "radiation sterilization" which would imply high dose treatments, with the resulting product being shelf-stable.

*Radurization* refers to "radiation pasteurization" low-dose treatments, where the intent is to extend a product's shelf life.

*Radicidation* also is a low-dose "radiation pasteurization" treatment, but with the specific intent being the elimination of a particular pathogen.

*Picowaved* is a term used to label foods treated with low-level ionizing radiation.

**X-Rays**

X-rays, gamma rays, and cathode rays are equally effective in sterilization for equal quantities of energy absorbed. X-rays and gamma rays have good penetration, while cathode rays have comparatively poor penetration. The greatest drawback at present to the use of x-rays in food preservation is the low efficiency and consequent high cost of their production, for only about 3 to 5 percent of the electron energy applied is used in the production of x-rays. For this reason, most recent research has concentrated on the application of gamma rays and cathode rays.

**Gamma Rays and Cathode Rays**

Since these two types of rays are equally effective in sterilization for equal quantities of energy absorbed and apparently produce similar changes in the food being treated, they will be discussed together and compared where possible.

***Sources***

Chief sources of gamma rays are (1) radioactive fission products of uranium and cobalt, (2) the coolant circulated in nuclear reactors, and (3) other fuel el-ements used to operate a nuclear reactor. Cathode rays usually are accelerated by special electrical devices. The greater this acceleration (i.e., the more meV), the deeper the penetration into the food.

***Penetration***

Gamma rays have good penetration, but their effectiveness decreases exponentially with depth. They have been reported to be effective up to 20 cm in most foods, but this depth will depend on the time of exposure. Cathode rays, on the other hand, have poor penetration, being effective at only about 0.5 cm per meV when "cross

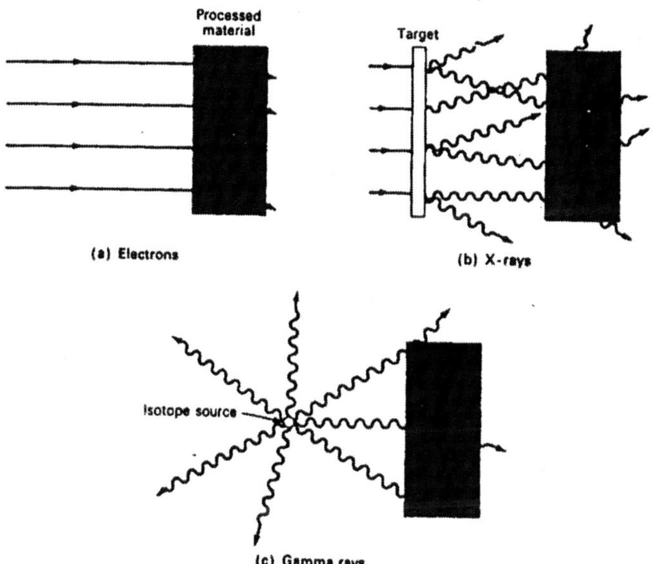


*Fig. 4.7. The three basic techniques for radiation processing—interactions of electrons, X-rays, and gamma rays in the medium.*

firing," that is, irradiation from opposite sides, is employed. The absorption dose level in a material is not a uniformly decreasing fraction with depth but rather builds up to a maximum at a depth equal to about one-third of the total penetration and then decreases to zero.

***Efficiency***

Because cathode rays are directional, they can be made to hit the food and therefore are used with greater efficiency than are gamma rays, which are constantly emitted in all directions from the radioactive sources. Various estimates of the maximal efficiency of utilization of cathode rays range between 40 and 80 percent, depending on the shape of the irradiated material, but only a maximum of 10 to 25 percent utilization efficiency is estimated for gamma rays. Radioactive sources of gamma rays decay steadily and hence weaken with time.

***Safety***

The use of cathode rays presents fewer health problems than the use of gamma rays, since cathode rays are directional and less penetrating, can be turned off for repair or maintenance work, and present no hazard of radioactive materials after a fire, explosion, or other catastrophe. Gamma rays are emitted in all directions, are penetrating, are continuously emitted, and come from radioactive sources. Gamma rays require more shielding to protect workers. Tests thus far on animals and human volunteers have not indicated any ill effects from the eating of irradiated foods.

**Effects on Microorganisms**

The bactericidal efficacy of a given dose of irradiation depends on the following:

1. *The kind and species of organism.* The importance of this factor is illustrated by the data in Table 4.22.
2. *The numbers of organisms (or spores) originally present.* The more organisms there are, the less effective a given dose will be.

**Table 4.22. Approximate killing doses of ionizing radiations in kilograys.**

| *Organism* | *Approximate lethal dose* | *Organism* | *Approximate lethal dose* |
|---|---|---|---|
| Humans | 0.0056-0.0075 | Bacteria (cells of saprophytes) | |
| Insects | .22-93 | Gram-negative | |
| Viruses | 10-40 | Escherichia coli | 1.0-2.3 |
| Yeasts (fermentative) | 4-9 | Pseudomonas aeruginosa | 1.6-2.3 |
| Saccharomyces corevisiae | 5 | Pseudomonas fluorescens | 1.2-2.3 |
| Torula cremoris | 4.7 | Enterobacter aerogenes | 1.4-1.8 |
| Yeasts (film) | 3.7-18 | Gram-positive | |
| Hansenula sp. | 4.7 | Lactobacillus spp. | 0.23-0.38 |
| Candida krusei | 11.6 | Streptococcus faecalis | 1.7-8.8 |
| Molds (with spores) | 1.3-11 | Leuconostoc dextranicum | 0.9 |
| Penicillium spp. | 1.4-2.5 | Sarcina lutea | 3.7 |
| Aspergillus spp. | 1.4-3.7 | Bacterial spores | 3.1-37 |
| Rhizopus sp. | 10 | Bacillus subtilis | 12-18 |
| Fusarium sp. | 2.5 | Bacillus coagulans | 10 |
| Bacteria (cells of pathogens) | | Clostridium botulinum (A) | 19-37 |
| Mycobacterium tuberculosis | 1.4 | Clostridium botulinum (E) | 15-18 |
| Staphylococcus aureus | 1.4-7.0 | Clostridium perfringens | 3.1 |
| *Corynebacterium diphtheriae* | 4.2 | Putrefactive anaerobe 3679 | 23-50 |
| Salmonella spp. | 3.7-4.8 | *Bacillus stearothermophilus* | 10-17 |

3. *The composition of the food.* Some constituents, e.g., proteins, catalase, and reducing substances (nitrites, sulfites, and sulfhydryl compounds), may be protective. Compounds that combine with the

SH groups would be sensitizing. Products of ionization may be harmful to the organisms.

4. *The presence or absence of oxygen.* The effect of free oxygen varies with the organism, ranging from no effect to sensitization of the organism. Undesirable "side reactions," which will be discussed subsequently, are likely to be intensified in the presence of oxygen and to be less frequent in a vacuum or an atmosphere of nitrogen.
5. *The physical state of the food during irradiation.* Both moisture content and temperature affect different organisms in different ways.
6. *The condition of the organisms.* Age, temperature of growth and sporulation, and state-vegetative or spore-may affect the sensitivity of the organisms. These factors have been discussed previously in connection with other methods of processing foods.

The type of irradiation and, within limits, the pH of the food seem to have little influence on the dose needed to inactivate the organisms.

It has been stated by some workers that the resistance of a given species of microorganism to ionizing radiations parallels in general its resistance to conventional heat processing, although there are notable exceptions. Spores of *Clostridium botulinum*, for example, have been found to be more resistant to gamma rays than are spores of a flat sour bacterium and a thermophilic anaerobe, although the latter two are more heat-resistant. Table 4.22 summarizes reports from various sources on the approximate dosages of radiation necessary to kill various types of microorganisms. These figures will vary with the conditions listed in the preceding paragraph. It is to be noted, however, that (1) humans are much more sensitive to radiations than are microrganisms, (2) bacterial spores are considerably more resistant than are vegetative cells, (3) gram-negative bacteria are, in general, less resistant than are gram-positive ones, and (4) yeasts and molds vary considerably, but some are more resistant than most bacteria. For example, *Candida krusei* is as resistant as many bacterial spores. Certain microorganisms are much more resistant than anticipated. For example, a radiation-resistant micrococcus resembling *Micrococcus roseus* has been found in irradiated meat. *Microbacterium* species in meat have been found to be especially resistant. There is, then, the possibility that resistant strains may be isolated from low-level-irradiated foods.

Preliminary gamma irradiation of bacterial spores has been found to make them more sensitive to heat, but preliminary heat shocking does not affect the lethal action of subsequent gamma irradiation. Previous ultrasonic treatment of organisms sensitizes them to radiation.

It is supposed that irradiated microorganisms are destroyed by passage of an ionizing particle or quantum of energy through, or in close proximity to, a sensitive portion of the cell, causing a direct "hit" on this target, ionization in this sensitive region, and subsequent death of the organism. It is assumed also that much of the germicidal effect results from ionization of the surroundings, especially of water, to yield free radicals, some of which may be oxidizing or reducing and therefore helpful in the destruction of the organisms. Irradiation also may cause mutations in the organisms present.

**Effects on Foods**

Radiation doses heavy enough to affect sterilization have been found to produce undesirable "side reactions," or secondary changes, in many kinds of foods, causing undesirable colors, odors, tastes, or even physical properties.

Some of the changes produced in foods by sterilizing doses of radiation inelude (1) in meat, a rise in pH, destruction of glutathione, and an increase in carbonyl compounds, hydrogen sulfide, and methyl mercaptan, (2) in fats and lipids, destruction of natural antioxidants, oxidation followed by partial polymerization, merization, and increase in carbonyl compounds, and (3) in vitamins, reduction in most foods of levels of thiamine, pyridoxine, and vitamins $B_{12}$, C, D, E, and K; riboflavin and niacin are fairly stable. The lower the dosage of irradation, of course, the less frequent the undesirable effects on the food. Destruction of many of the food enzymes requires five to ten times the dosage of rays needed to kill all the microorganisms. Enzyme action may continue after all microorganisms have been destroyed unless a special blanching treatment has preceded irradiation.

The chief effect on the healthfulness of foods is the destruction of vitamins. However, the general nutrition of an irradiated food would be as good as that of a food processed by other means to achieve the same shelf stability. There is no indication of production of radioactivity with electron beams below 11 meV or with gamma rays from cobalt 60.

**Applications**

Currently food irradiation has been approved only in a very limited way in the United States. Low-level irradiation (1 kiloGray) can be used on fresh fruits and vegetables to kill insects and to inhibit spoilage. Dry or dehydrated vegetables (herbs and spices) can be irradiated at up to 30 kiloGray to kill insectsand bacteria. Additionally, processes for irradiating pork (trichinae), wheat (insects), and potatoes (sprout formation) have been approved. However, renewed interest and regulatory proposals will probably result in expanded use in this country. Close to thirty countries now have approved some form of food irradiation. The specific applications of food irradiation are summarized in Table 4.23.

**Table 4.23. Applications of food Irradiation.**

| *Type of food* | *Radiation dose In kiloGrays* | *Effect of treatment* |
|---|---|---|
| Meat, poultry, fish, shellfish, some vegetables, baked goods, prepared foods | 20-70 | Sterilization. Treated product can be stored at room tempera ture without spoilage. Treated product is safe for hospital patients who require microbiologically sterile diets |
| Spices and other seasonings | 8-30 | Reduces number of microorganisms and insects. Replaces chemicals used for this purpose |
| Meat, poultry, fish | 1-10 | Delays spoilage by reducing the number of microorganisms in the fresh, refrigerated product. Kills some types of food poisoning bacteria |
| Strawberries and some other fruits | 1-4 | Extends shelf life by delaying mold growth |
| Grain, fruit, vegetables, and other foods subject to insect infestation | 0.1-1 | Kills insects or prevents them from reproducing. Could partially replace fumigants used for this purpose |
| Bananas, avocados, mangos, | 0.25-0.35 | Delays ripening papayas, guavas, and certain other noncitrus fruits |
| Potatoes, onions, garlic | 0.05-0.15 | Inhibits sprouting |
| Pork | 0.08-0.15 | Inactivates trichinae |
| Grain, dF hydrated vegetables, other foods | Various doses | Desirable physical and chemical changes |

## Microwave Processing

Microwave heating and processing of foods is becoming increasingly popular, particularly at the consumer level. Microwaves are electromagnetic waves between infrared and radio waves. Specific frequencies are usually at either 915 megacycles or 2,450 megacycles. The energy or heat produced by microwaves as they pass through a food is a result of the extremely rapid oscillation of the food molecules in an attempt to align themselves with the electromagnetic field being produced. This rapid oscillation, or intermolecule friction, generates heat. The preservative effect of microwaves or the bactericidal effect produced is really a function of the heat that is generated. In other words, the microwaves themselves do not result in any inactivation of foodborne microorganisms; rather, it is the heat produced by the excitation of food molecules that actually results in microbial destruction.

# 5

# YEAST IN INDUSTRIES

Yeasts are eucaryotic organisms that exist in nature predominantly as single cells. Traditionally, they have played an important role for man, being used for thousands of years in bread-making, brewing, and making certain foods palatable and nutritious. The commercial importance of yeast has grown considerably over the past few decades, and they are now being used in a variety of fermentative processes for the synthesis of simple sugars and ammonium nitrogen as well as certain fats, vitamins, and proteins. Although there are about 350 recognized species of yeast, only a relatively small number are important commercially, notably members of the genera *Saccharomyces* and *Candida*. Species of *Saccharomyces*, such as *S. cerevisiae* and *S. uvarum* (which is sometimes referred to as *S. carlsbergensis*) are used in the manufacture of breads, wines and beers, while some species of *Candida* are sources of animal food and fodder. One species in particular, *S. cerevisiae*, has received an enormous amount of attention from generations of biologists. This is not so much due to its commercial importance, but rather because it is amenable to most of the cultural techniques and genetic manipulations used with laboratory bacteria. It has been extremely well-characterized genetically and defined mutations in most biochemical pathways have been established.

The genetic engineering of yeast using recombinant DNA techniques has been made possible by the development, recently, of a transformation system and the construction of DNA cloning vectors which mediate the introduction of DNA into yeast cells. These technological advances, combined with classical genetic studies, have provided a wealth of knowledge of yeast molecular biology. The ability to introduce individual genes into yeast and apply selection by complementation to identify a clone carrying a particular gene now permits the isolation of virtually any yeast gene provided that a suitable recipient yeast strain is available. Furthermore, gene cloning in yeast is likely to be of great industrial importance, not only for the improvement of existing strains used in fermentation processes, but also for the bulk production of commercially important proteins not selectable in or compatible with bacterial systems.

## Transformation

Gene cloning is a method whereby DNA molecules are joined together *in vitro* and introduced into living cells where they are able to be maintained and replicated. An integral part of this method is transformation which is a process whereby cells take up and maintain exogenously-supplied DNA and express genes on the DNA to alter the phenotype of the recipient cells. In common with many organisms, yeast does not have a natural system for DNA uptake and all procedures for yeast transformation rely on artificial methods to introduce the products of *in vitro* DNA manipulation into cells. They are based on the effects of polyethylene glycol on the surface of spheroplasts, and on the reasonably efficient regeneration of normal cells from spheroplasts.

### Procedures

Yeast cells are surrounded by a thick cell wall which is composed mostly of the polysaccharides mannan and glucan, with some protein and lipid components, and a small amount of chitin. The preparation of spheroplasts involves enzymatically removing the cell wall with mixtures of β-glucanases. Some commercial preparations of this mixture that are widely used for this purpose are Glusulase and Helicase, both crude extracts of snail gut. Partially pure preparations, such as Zymolase and Lyticase are now widely used. However, Glusulase and Helicase have the advantage that overdigestion with these enzyme mixtures is virtually impossible, whereas Zymolase and Lyticase treatments must be carefully monitored because long exposure to the enzymes seriously disrupts cell metabolism leading to poor regeneration of spheroplasts. Prior treatment of yeast cells with a reducing agent, such as 2-mercaptoethanol or dithiothreitol, is sometimes used to sensitize the cell wall to subsequent enzymatic degradation by reducing disulfide bridges in the protein components of the wall thus making the glycosidic bonds available for cleavage.

**Table 5.1. A Procedure for yeast transformation.**

| *Materials* | |
|---|---|
| SED | SOS |
| 1 *M* sorbitol | 10 ml 2 *M* sorbitol |
| 25 *mM* EDT A, pH 8.0 | 6.7 ml yeast nutrient medium |
| 50 *mM* dithiothreitol | 0.13 ml 1 *M* calcium chloride |
| | 27 μl 1% solution of amino acid to be selected |
| SCE | 3.17 ml water |
| 1 *M* sorbitol | Regeneration agar |
| 0.1 *M* sodium citrate, pH 5.8 | 182 g sorbitol |
| 10 *mM* EDTA | 20 g agar |
| STC | 6.7 g Difco yeast nitrogen base |
| 1 *M* sorbitol | 20 g glucose |
| 10 *mM* calcium chloride | amino acids as required (to 40 μg/ml) |
| 10 *mM* Tris-HCI, pH 7.5 | water to 1 liter |
| PEG | All solutions are sterilized. |
| 20% (w/v) polyethylene glycol, M.W. 4000 | |
| 10 *mM* calcium chloride | |
| 10 *mM* Tris-HCI, pH 7.5 | |

Procedure

1. Inoculate a flask containing 100 ml of yeast nutrient medium with an overnight culture of yeast to give a starting O.D.600 of 0.1 (equivalent to 5 × $10^7$ cells/ml).
2. Grow the cells with shaking at 30°C until the culture has reached an O.D.600 of 0.4 (about 2 × $10^7$ cells/ml).
3. Harvest the cells by centrifugation at 5,000 rpm for 5 min. Wash the cell pellet once with 10 ml of 1 *M* sorbitol.
4. Resuspend the cells in 10 ml of SED. Incubate at 30°C for 10 min with occasional gentle shaking.
5. Collect the cells by low-speed centrifugation. Resuspend the cells in 10 ml of SCE. Add 0.1 ml of Glusulase and incubate at 30°C for 20 to 30 min, occasionally gently shaking the suspension.
6. Monitor for speroplast formation by diluting 10 ml of the cell suspension in a drop of 5% (w/v) sodium lauryl sulfate on a microscope slide and observing "ghosts" under the phase contrast microscope.
7. Collect the spheroplasts by spinning at 3,000 rpm for 4 min. Wash twice with 10 ml each of 1 *M* sorbitol.

8. Resuspend in 1 ml of STC. Pipette 0.2 ml aliquots of the suspension into 10 ml Falcon tubes.
9. Add 1 to 5 micrograms of DNA. Incubate the tubes at room temperature for 5 min.
10. Add 2 ml of PEG to each tube and leave at room temperature for a further 10 min.
11. Centrifuge the tubes at 3,000 rpm for 4 min. Resuspend the spheroplasts in 0.5 ml of SOS and incubate at 30°C for 30 min. At this point the cells may be plated or held at 4°C for a few days.
12. Add 6 ml of regeneration agar (kept at 45°C) to each tube and pour the contents immediately onto pre-warmed plates. Incubate the plates at 30°C for 2 to 7 days to allow transformants to grow.

---

Spheroplasts are maintained in an hypertonic medium provided by 1 *M* sorbitol or 0.6 *M* KCI. Uptake of naked DNA is promoted by co-precipitating calcium-treated spheroplasts and transforming DNA with polyethylene glycol. Regeneration of the cell wall and selection of transformed colonies is done simultaneously by embedding the spheroplasts in a solid matrix composed of 3 percent agar and selective medium which permits only the growth of transformed cells. Colonies develop in the agar within two to seven days depending upon the particular DNA and recipient strain used in the transformation.

A few variations have been developed on this basic method. In one procedure, *Escherichia coli* protoplasts carrying recombinant plasmids can be fused directly to yeast spheroplasts and yeast transformants selected by plating the entire fusion mixture in yeast regeneration agar. This technique, known as yeast transfusion, is relatively inefficient compared to conventional yeast transformation but does have the advantage of eliminating the time and labor spent in the isolation of recombinant plasmid DNA from *E. coli*.

An even more rapid procedure for yeast transformation has been developed recently that completely eliminates the need to make spheroplasts, an often tedious and time-consuming procedure. Intact yeast cells are made competent by treating them with alkali metal ions such as lithium and rubidium salts. Incubation of competent cells with polyethylene glycol promotes the uptake of naked DNA molecules and transformants can be grown directly on selective agar plates without using regeneration agar. The transformation efficiency obtained with this method is generally lower than that obtained with conventional spheroplast techniques.

**Table 5.2. A Procedure for yeast transformation of intact yeast cells.**

| Materials | | |
|---|---|---|
| TE | LC | PEG |
| 10 *mM* tris-HCl, pH 7.5 | 0.1 *M* lithium chloride | 50% (w/v) polyethylene glycol, M.W. 4000, in water |
| 1 *mM* EDTA | 10 *mM* tris-HCl, pH 7.5 | |
| | 1 *mM* EDTA | |

All solutions are filter-sterilized.

Procedure

1. Inoculate 100 ml of yeast nutrient medium with an overnight culture of yeast to give a starting O.D. 600 of 0.1. Grow the cells at 30°C with shaking to an O.D. 600 of 0.4.
2. Harvest the cells by centrifugation and wash once with 10 ml of TE.
3. Resuspend the cell pellet in 1.5 ml of LC. Incubate the suspension at 30°C for 60 min.
4. Pipette 0.2 ml of the suspensicm into 10 ml Falcon tubes. Add 1 to 10 $\mu$g of DNA. Incubate at 30°C for 30 min.
5. Add 1.2 ml of PEG to each tube. Incubate at 30°C for a further 50 min.
6. Heat shock the suspension at 42°C for 5 min. Pellet the cells by low-speed centrifugation and wash twice with 10 ml of sterile water.
7. Finally, resuspend the cells in 0.2 ml of sterile water and plate directly onto selective agar plates. Incubate the plates at 30°C for a few days to allow transformants to grow.

### Genetic Markers for Yeast Transformation

Selection of successful transform ants is based on the ability of genes carried on the transforming DNA to complement auxotrophic mutations in the recipient strain. The first yeast genes characterized were responsible for the synthesis of enzymes involved in amino acid biosynthesis, and were isolated from yeast genome libraries by their ability to complement mutations in *E. coli*. Subsequently, these genes have been used extensively in the construction of yeast cloning vectors. Several host-vector systems have been developed by combining these vectors with the corresponding auxotrophic yeast strain as recipient. However, since selective pressure for successful transformants depends upon the reversion frequencies of these mutations (which is usually of the order of $10^{-6}$ to $10^{-7}$), the appearance of a large number of revertants may obscure the identification of transformed cells, especially when transformation has been inefficient. This problem has been largely overcome by making stable mutations for most of the commonly used auxotrophic markers. These include introducing double or triple mutations into a gene, or by the *in vitro* construction of deletions inserted into the yeast genome by replacement of the wild-type gene.

Constructing recipient strains with suitable genotypes to be used in transformation is difficult and time-consuming. Antibiotic resistance determinants carried on the transforming DNA should be generally useful as selectable markers and could be used with any sensitive yeast strain as recipient. However, although some antibiotic resistance genes of bacterial origin are expressed in yeast, they cannot be used as selectable markers in yeast transformation because yeast cells are naturally resistant to most of these antibiotics. An exception is G418, an aminoglycoside antibiotic which inhibits the growth of a wide range of procaryotic and eucaryotic organisms, including yeast. Resistance to G418 is encoded by a bacterial transposable element, *Tn*601, and yeast cells transformed with the element are resistant to high levels of the antibiotic. The use of G418 for yeast transformation selection, by eliminating the need for specially constructed recipient strains, should be particularly valuable in the genetic engineering of industrial yeast strains.

## Yeast Cloning Vectors

Nearly all systems for isolating and analyzing gene sequences to be cloned in yeast require the initial isolation of recombinant DNA molecules in *E. coli*. This gives the experimenter several practical advantages. *E. coli* offers a high efficiency of transformation, powerful hybridization screening methods for the detection of cloned DNA sequences, and good amplification of recombinant DNA molecules to be used in yeast transformation. The ability to maintain cloned genes in two alternative hosts also allows the experimenter several options carrying out manipulations on the cloned DNA by taking advantage of the particular properties of each host and its mutants. The vectors currently used in gene cloning in yeast are therefore hybrids composed of bacterial plasmid or bacteriophage DNA and yeast DNA sequences.

The particular purposes of cloning experiments generally dictate which type of vector system is used. For instance, if a small segment of DNA is to be purified for the genetic manipulation of yeast, then plasmid systems are generally employed. Plasmid vectors offer the advantages of high purification of cloned DNA sequences and simplicity of manipulation of plasmid DNA, and are by far the most popular choice of vectors for gene cloning in yeast. However, if random DNA fragments are to be screened by physical methods in which completeness of representation and large-sized inserts are desired, then other types of cloning vectors must be used, the most generally useful being hybrid plasmid-bacteriophage vectors called cosmids.

The different modes by which DNA can be maintained in yeast following transformation has led to the development of two main classes of cloning vectors: integrating vectors and autonomously-replicating vectors. Each type of vector has different characteristics and uses, but all have the following properties in common:

1. They contain suitable genetic markers that can be selected in yeast recipient strains defective for that gene. For some genes (such as *LEU2* and *TRP1*) selection can also be carried out in suitable *E. coli* hosts as well as yeast. However, most of the vectors in general use carry one or more antibiotic resistance genes, and selection for the presence of these markers can be carried out in virtually any *E. coli* strain.
2. All vectors carry an origin of replication for maintenance in *E. coli* cells which often results in a high copy number.
3. They contain single sites for one or more restriction endonucleases suitable for the cloning of foreign DNA fragments.

**Integrating Vectors**

Yeast integrating vectors (often referred to as YIp) consist of a bacterial cloning vector which can be either a plasmid, bacteriophage, or cosmid, and a suitable yeast gene which can be used for the selection of transformanl cells. Transformation is accomplished by integration of the vector into the recipient yeast genome and is mediated by recombination between the yeast DNA sequence on the vector and the homologous DNA sequence in the nuclear DNA of the recipient cell. The transformation efficiencies obtained with these vectors are very low, being of the order of about 1 to 10 transformants per microgram of vector DNA (equivalent to 1 transformant per $10^6$ to $10^7$ viable, regenerated cells). The low number of transformants obtained does not normally present a major problem to the experimenter, since high yields of vector DNA can be obtained by initial amplification in *E. coli*. Much higher transformation frequencies are obtained by linearizing the vector DNA through cleavage in the yeast moiety. This also has the effect of directing the integration of the transforming plasmid at a

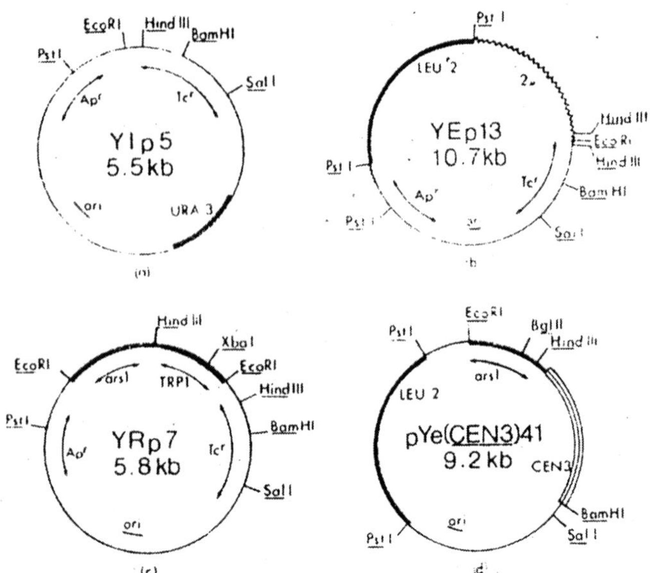


*Fig. 5.1. Physical maps of yeast cloning plasmids. $Ap^r$ and $Tc^r$ are resistance determinants for ampicillin and tetracycline, respectively; ori is the bacterial origin of replication.*

specific site on the chromosome. The free DNA ends are apparently recombinogenic and act to increase the efficiency of integration and, thereby, transformation. The efficiency of transformation may be increased still further by cotransforming yeast with a YIp vector and an extrachromosomally-replicating vector. If the vectors have DNA sequences in common, recombination can take place between them leading to a single extrachromosomally-replicating molecule. This feature of cotransformation has other practical advantages. Only one of the two vectors needs to have a yeast selectable marker and since recombination can take place between shared bacterial sequences it can provide a convenient way of introducing genes into yeast which have been cloned on a purely bacterial vector.

The transformed phenotype obtained with YIp vectors is very stable, the rate of segregation of integrated vectors being less than 1 percent per generation of growth in non-selective medium. The segregation frequency seems to be correlated with the amount of homology between recipient and vector DNA, and extensive homology results in higher transformation as well as higher segregation frequencies. Normally, there is only a single copy of the transforming vector DNA in each

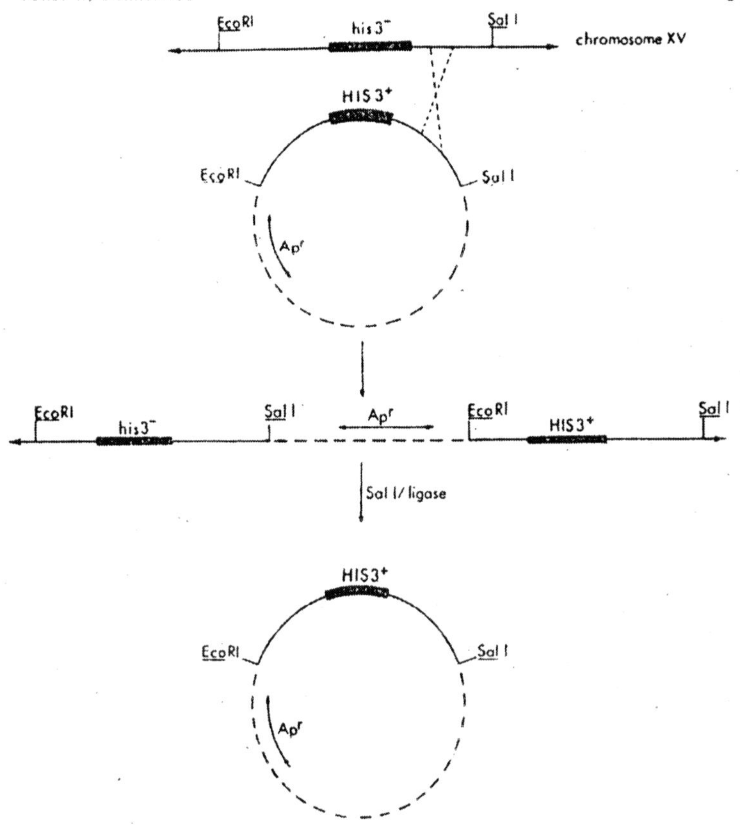


*Fig. 5.2. A scheme illustrating the integration of a YIp vector into yeast nuclear DNA and recovery of the YIp DNA.*

stably transformed cell, the integrated vector replicating with nuclear DNA and being inherited in the same manner as any regular chromosomal gene.

One major problem associated with integrating vectors is the tedious procedure needed to recover them from transformed cells. It is necessary to prepare total DNA from yeast transformants and digest it with a restriction endonuclease which cuts the vector DNA once within or at the boundary of the yeast DNA insert. The mixture of DNA fragments is then ligated at low DNA concentration and the recircularized vector recovered by transforming *E. coli* cells and selecting for an antibiotic resistance gene on the vector DNA. Depending upon the location of the integrated vector DNA with respect to the chromosomal marker gene and the sites of restriction endonuclease cleavage, either the mutant allele (originally from the recipient genome) or the wild-type allele (originally from the vector) of the yeast marker gene will be recovered.

Yeast integrating vectors are exceptionally useful tools for genetic analysis and gene manipulation in yeast. The specific integration and excision mechanism of YIp vectors has been used to map cloned genes and provides the possibility of mutating yeast structural genes in a defined manner. For instance, they can be used for the construction of strains carrying gene duplications or substitutions by virtue of their integration into nuclear DNA. Sophisticated methods have been developed whereby cloned genes, mutated using *in vitro* techniques, can be recombined into recipient yeast DNA to replace the wild-type allele. These methods should have potentially wide applications for constructing yeast strains carrying fully characterized mutations in virtually any yeast gene.

**Extrachromosomally-Replicating Vectors**

The ability of cloning vectors to replicate extrachromosomally in yeast is conferred by special yeast sequences derived from either the endogenous yeast 2-micron plasmid or nuclear DNA. Those vectors that include a segment of 2-micron plasmid DNA for maintenance are called yeast episomal plasmids (YEp); vectors that contain chromosomal replicators or autonomously-replicating sequences (*ars*) are called yeast replicating plasmids (YRp); and, finally, vectors that contain a functional centromere are referred to as yeast centromere plasmids (YCp).

***Yeast episomal plasmids***

Many, if not all strains of *S. cerevisiae*, contain an endogenous plasmid often referred to as the 2-micron plasmid, or *Scp*1, that replicates independently of the chromosomes. 50 to 100 copies of the plasmid are present in each cell and it appears to replicate under cell nuclear control. The entire nucleotide sequence of the plasmid has now been established and reveals two inverted repeat sequences, 598 base-pairs in length. Recombination between the inverted repeat sequences results in the coexistince of two forms of the plasmid molecule which differ in the respective orientation of the two non-repeated sequences.

Cloning vectors that contain certain segments of this plasmid also behave like autonomously-replicating units and transform yeast with a high efficiency ($10^3$ to $10^5$ transformants per microgram of DNA) that is several orders of magnitude greater than integrating vectors. The copy number of the plasmids after transformation is quite high, ranging from 25 to 100 copies per cell, which may be important where a high level of expression of a cloned gene is required. They can be recovered as plasmids from yeast cells, but the stability of the transformed phenotpye varies depending upon the size of the transforming plasmid and which segment of the 2-micron plasmid is present in the vector. The smaller and more versatile cloning vectors, such as YEp13 contain only a small portion of the 2-micron plasmid which provides high transformation efficiency, multiple copies of the vector and relatively stable maintenance in yeast cells.

Some disadvantages, however, are apparent with these cloning vectors. Sequence rearrangements, resulting in complex recombinant molecules, sometimes occur at low frequencies and are due to homologous recombination between 2-micron DNA on the vector and the resident 2-micron plasmid in recipeint cells. Also, instability of the transforming vector seems to increase with additionally inserted DNA and may be due to replicative competition with the resident plasmid. The obvious solution to these problems is to use 2-micron plasmid-free mutants as host cells. Vectors containing the entire 2-micron plasmid transform these strains with high efficiency and transformants are extremely stable. However, it appears that most YEp vectors which are relatively stable in yeast strains containing the 2-micron plasmid are rapidly lost in plasmid-free strains. This is presumably due to the absence on these vectors of certain 2-micron sequences required for efficient replication.

***Yeast replicating vectors***

Yeast replicating vectors contain segments of yeast chromosomal DNA carrying a biochemical genetic marker closely associated with an autonomously-replicating sequence which functions as an origin or replication in yeast. Similarly to yeast episomal vectors, they transform yeast with high efficiency ($10^2$ to $10^4$ transformants per microgram of DNA), but the yeast transformants obtained contain only one or two copies of the plasmid per cell. This may be advantageous for cloning genes with deleterious effects. A major disadvantage of these vectors is their high mitotic and meiotic instability in transformed cells, the plasmid being lost at a rate greater than 1 percent per generation of growth in nonselective medium. Sometimes, however, rare stable transformants can be obtained in which the transforming vector has integrated at a homologous site on the yeast genome in an identical manner to yeast integrating vectors. This property of yeast replicating plasmids may be useful since it combines a high frequency of transformation with the possibility of stable integration through homologous recombination into yeast nuclear DNA.

***Yeast centromere vectors***

Artificial minichromosomes have been constructed recently which consist of a yeast vector containing a chromosomal or plasmid origin of replication and DNA sequences apparently derived from yeast centromeres. These plasmids behave in mitosis and meiosis as stably inherited additional linkage groups. More so than yeast replicating vectors, these plasmids represent potentially useful genetic systems, combining the advantages of efficient transformation with stable propagation of a single gene dosage carried on the vector.

## DNA Cloning in Yeast Plasmids

The procedures for introducing foreign DNA fragments into yeast plasmid vectors to generate viable recombinant molecules are straightforward in principle, but in practice care is taken in order to minimize the subsequent effort required to identify and characterize recombinant clones. Insertion of foreign DNA is usually by simple ligation of plasmid DNA cut with a restriction endonuclease in the presence of excess foreign DNA cut with the same enzyme. In a ligation mixture, those molecular species that are most effective in *E. coli* transformation and subsequent stable plasmid propagation are circular molecules. Circle formation is optimum when the concentration of DNA fragment ends is low, the only major undesirable product of the reaction being circularized plasmid containing no foreign DNA insert. Recircularization of plasmid DNA may be limited to some extent by adjusting the relative concentrations of vector and foreign DNAs during ligation. However, this problem is dealt with most efficiently by using either biochemical methods, such as treatment of vector DNA with an alkaline phosphatase to remove 5'-phosphoryl groups, to prevent the recircularization reaction, or genetical methods which involve the identification or selection of plasmids that contain foreign DNA inserts.

Cloning vectors able to replicate autonomously in both *E. coli* and yeast are generally referred to as "shuttle vectors." They are easy to recover from yeast by a simple DNA extraction followed by transformation of *E. coli*. This makes it possible to transfer, on a routine basis, genetic information back and forth between *E. coli* and yeast. Nearly all of the widely used plasmid vectors are based on the bacterial plasmid, pBR322, which carries genes for resistance to ampicillin and tetracycline. A typical vector of this type, YEp13, can be selected in *E. coli* for these antibiotic resistances or by complementation of *leu*B mutations, and selected in yeast for complementation of *LEU2* lesions. There are several restriction endonuclease sites available for cloning, the most useful being the *Bam*H1 cleavage site which is situated in the tetracycline resistance gene; insertion of foreign DNA at this site inactivates this gene and provides a method for distinguishing recombinant molecules from recircularized plasmid DNA.

Some cloning vectors have been modified to accept DNA fragments made with almost any restriction endonuclease producing cohesive ends, and others have been constructed for cloning genome fragments at random using partial digests with restriction endonucleases of low specificity that produce cohesive ends complementary to those generated by an enzyme that cleaves the vector DNA just once (for example, *Sau*3A for *Bam*HI vectors). These vectors are generally used to construct and amplify DNA libraries from small genomes in *E. coli*. The library can be transferred to yeast and the high frequency of transformation obtained facilitates the screening of large numbers of transform ants for specific gene sequences. Individual transformants are then returned to *E. coli* for preparative isolation and genetic manipulation. These methods have been widely used not only to isolate and characterize many yeast genes which cannot be isolated by cloning in bacterial systems, but also some genes from other eucaryotic organisms that are able to complement certain yeast mutations.

## DNA Cloning in Yeast Cosmids

Yeast cosmids have been designed for the convenient cloning of large DNA fragments in *E. coli* and their transfer to yeast. The recovery of recombinant clones carrying large DNA inserts is sufficiently high to establish gene libraries of complex eucaryotic genomes from microgram amounts of DNA. Yeast cosmids consist of (a) bacterial plasmid DNA sequences for replication and selection in *E. coli*; (b) yeast 2-micron plasmid DNA sequences or chromosomal replicators and yeast markers necessary for maintenance and selection in yeast; and (c) the cohesive end fragments (*cos*) of bacteriophage lambda which allow packaging of large recombinant molecules into lambda phage heads. Restriction and ligation of large DNA fragmenls into restriction endonuclease-cleaved cosmid DNA results in long concatenated DNA structures that resemble the natural DNA substrate of the lambda packaging system. Hybrid concatamers with *cos* sites

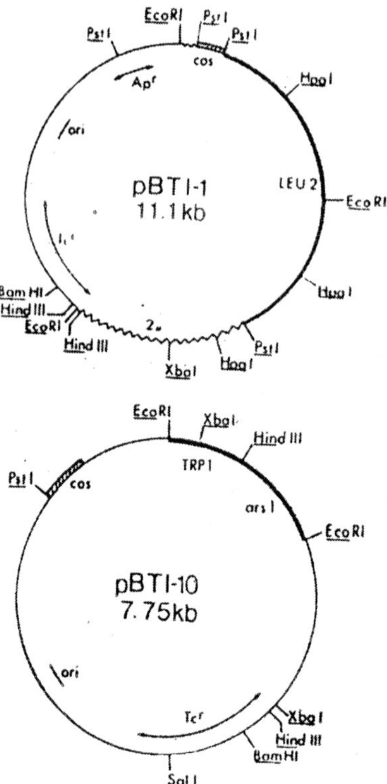


*Fig. 5.3. Physical maps of yeast cosmids, pBTI-1 and pBTI-10. $Ap^r$ and $Tc^r$ are resistance determinants for ampicillin and tetracycline, respectively; ori denotes the bacterial origin of replication.*

situated approximately 36 to 50 kilobase-pairs apart, and in the same orientation, are packaged using an *in vitro* system into lambda particles which can then be transduced into *E. coli* with high efficiency. The small size of yeast cosmids (about 10 kilo base-pairs) and the size selection imposed by the packaging reaction, allows the cloning of DNA fragments 25 to 40 kilobase-pairs in size.

The use of restriction endonucleases with six-base-pair recognition sites to cleave genomic DNA is usually avoided because they can lead to the exclusion of very large or very small DNA fragments during cloning. Instead, foreign DNA is partially digested with a restriction endonuclease recognizing a four-base-pair sequence that cuts DNA more frequently to produce a more random fragment population. Examples are *Sau*3A and *Mbo*I which cleave DNA to produce cohesive ends complementary to those generated by *Bam*HI which cleaves most vector DNAs just once. In order to ligate DNA fragments into the vector in an authentic gene arrangement without obtaining a random assortment of insert fragments, the endonuclease-cleaved DNA is fractionated by ultracentrifugation on linear sucrose gradients to purify DNA fragments between 20 and 40 kilo base-pairs in size. Finally, the ligation reaction is carried out at high DNA concentrations to ensure the formation of concatameric DNA.

Transduction procedures, in which an *in vitro* system is used to package recombinant cosmid DNA into infectious lambda phage particles, is an extremely efficient method for the transfer of large DNA molecules to *E. coli*. The *in vitro* packaging procedure uses a combination of extracts of two *E. coli* lysogens each of which is defective in a different step of lambda morphogenesis. In one lysogen which contains a prophage mutation in gene *A*, empty precursor particles accumulate within the bacterial cell upon induction. This mutation can be complemented by the addition of the missing *A* protein supplied by an induced lambda *E* lysogen. This mutation affects the synthesis of the main capsid protein; in its absence, all other head proteins are available in soluble form. Ligated DNA is added to a concentrated, lysed mixture of the two lysogens, and in the presence of ATP and polyamines, DNA is packaged, the phage head matures, and tails, which are made by both lysogens, are added. The process results in DNase-resistant, infectious virus particles which can be used to transduce sensitive *E. coli* cells like any *in vivo* produced bacteriophage. Transduced, recombinant cosmids circularize via the lambda cohesive ends within host cells, and since cosmids do not contain all the essential lambda phage genes, they do not go through a lytic cycle. Instead, recombinant molecules are maintained in their hosts as large, autonomously-replicating plasmids that can be amplified and used directly to transform appropriate recipient yeast cells.

The efficiency of yeast transformation by recombinant yeast cosmids seems to be unaffected by increase in size which is in sharp contrast to the strong dependence on plasmid size in *E. coli* transformation. They can be recovered from yeast either by transformation of *E. coli* with total yeast DNA or by *in vitro* packaging.

## Fusion Plasmids

Gene fusions have been very useful for investigating regulatory mechanisms in *E. coli*. In particular, gene fusions in the *lac*Z gene of *E. coli*, which specifies β-galactosidase synthesis, have been widely used because the enzyme retains its activity when up to 30 amino acid residues are removed from its amino terminus and replaced with other polypeptide residues. The ability of the *lac*Z gene to be functionally expressed in yeast has been widely used to investigate regulatory phenomena in yeast by fusing this gene to yeast genes. Yeast plasmids capable of replication in both *E. coli* and yeast and having selectable markers for both organisms have been constructed containing the *lac*Z gene which is missing the first few amino acid codons. In order to be expressed, *lac*Z must be ligated *in vitro* to a segment of DNA that encodes the regulatory signals and amino-terminal region of a gene in the correct reading frame. This is achieved by first manipulating the insert DNA with exonuclease *Bal*31 and synthetic oligonucleotide linkers which allow linkages in different translational reading frames at the restriction endonuclease cleavage site at the boundary of the *lac*Z gene. When introduced into yeast, successful fusions result in the expression of β-galactosidase enzyme activity which can be detected by including a chromogenic substrate in agar plates or by enzyme assay of yeast transformants. This kind of genetic tool has been used to study the mechanisms involved in the regulation of yeast genes and could also be used to indicate and measure the function of exogenously-derived promoters.

## Vectors for High Level Gene Expression

Many attempts have been made to isolate particular genomic sequences from other eucaryotic organisms by complementing defined yeast mutants affecting genes common to both organisms. This approach has met with limited success suggesting that there may be many differences in the mechanisms of gene expression between yeast and higher eucaryotes. Specialized yeast vectors are now available, however, which maximize the chances of functional expression of defined heterologous (i.e., non-yeast) eucaryotic genes in yeast by using endogenous yeast promoters. These expression vectors are composed of selectable yeast and bacterial genes, a yeast replication origin (plasmid or chromosomal), and transcription initiation and termination specifying sequences derived from highly expressed yeast genes that flank a restriction endonuclease cleavage site where foreign DNA is inserted. Powerful promoters are derived from the alcohol dehydrogenase I gene (*ADH*1) and the 3-phosphoglycerokinase gene (PGK).

In many gene expression experiments, the coding region of the gene to be expressed is joined to the yeast promoter fragment in such a way that the protein made has no amino terminal extension beyond that of the natural protein. Therefore, it is necessary to remove all

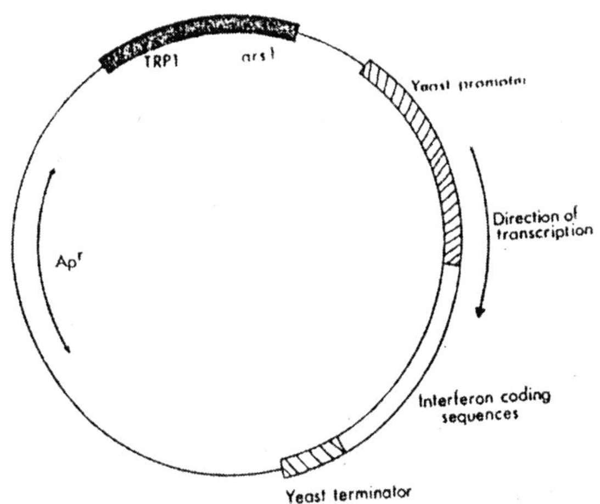


*Fig. 5.4. Schematic representation of a plasmid for the expression of an interferon gene in yeast.*

the protein coding units, including the initiation codon (ATG), from the yeast promoter fragment, and replace them with a natural or synthetic ATG initiator as part of the coding sequence to be expressed. This is usually done by making a series of deletions with exonuclease *Bal*31 that extend to varying distances into the untranslated leader mRNA of the endogenous promoter, then attaching synthetic linkers and cloning the deleted plasmids. By using this set of deletion plasmids, the coding region for any heterologous protein can be attached to the promoter region with a variety of spacings between the transcription and translation starting points. The different deletion plasmids mostly allow appreciable levels of heterologous gene expression indicating that a large part of the leader region may not be essential for the synthesis of a translatable mRNA.

Expression plasmids have been used for the synthesis in yeast of a number of proteins of biological and commercial importance which will be discussed in greater depth in the next section. There are, however, several requirements necessary for heterologous expression in yeast besides the need for a promoter sequence from yeast to initiate correct transcription. These are:

1. The coding sequences of the protein to be expressed must contain no intervening sequences of the type not processed in yeast. Thus, genomic sequences that are free of intervening sequences or complementary DNA (cDNA) synthesized *in vitro* from the mature mRNA of the gene are used.

2. The untranslated leader sequence just preceding the AUG initiation codon must have a specific sequence to support efficient translation initiation by yeast ribosomes. Highly expressed yeast genes contain only C, A, and an occasional T residue in this region, and the presence of G residues in the mRNA just before the AUG may adversely affect the initiation of translation.
3. The codon usage in all yeast genes which have been sequenced is biased towards a limited number of preferred codons. The general differences in codon usage patterns between different organisms may limit the expression of an heterologous gene in yeast whose codon usage patterns are inappropriate to the yeast intracellular tRNA population.

## Industrial Applications

The techniques described here have made possible the genetic engineering of yeast to obtain the expression of eucaryotic gene products having pharmaceutical and biological importance. There are, in fact, several features which make yeast an attractive organism for the large-scale production of biologically-active proteins. Yeast can be produced in large bulk using the technology of the brewing and fermentation industries, enabling a large quantity of the product to be isolated at low cost and in a system giving high purity. Yeast is capable of withstanding hydrostatic pressure and cells do not lyse after death even in cultures grown to high cell density. Furthermore, yeast will be very useful for the synthesis of glycoproteins since the addition of the inner core of carbohydrates occurs in yeast by the same mechanism as in animal cells.

The expression of several types of human interferon has been achieved in yeast using the expression vectors described above. The gene sequence coding for mature leucocyte interferon has been incorporated in extrachromosomally-replicating yeast vectors under the control of endogenous promoters derived from either the *ADH*1 gene or the *PGK* gene. The expression of $6 \times 10^6$ units of interferon per liter has been obtained, constituting a potential alternative method for the large-scale production of interferons. Much attention has been focused recently on the production of $\gamma$-interferon, which has been reported to have a potential antitumoral activity. Expression of the structural gene for human $\gamma$-interferon has now been achieved in several heterologous systems, including yeast. In this system, the mature interferon coding sequence was inserted behind the promoter for *PGK* when up to $10^8$ units of protein per liter were obtained, the product having identical properties to authentic human γ-interferon. In addition, the natural secretory pathway of yeast has now been altered to allow the natural secretion into the growth medium of interferons synthesized on expression vectors.

Other foreign genes which have been expressed in yeast using this system are the genes for human serum albumin, human growth hormone, epidermal growth factor and some insulin-like growth factors. Recently, techniques have been described whereby expression vectors have been used to synthesize the surface antigen of Hepatitis B virus in yeast. The protein produced in yeast is assembled into particles very similar to those secreted by human cells. This not only promises the large-scale production of the protein for possible use as a vaccine, but also illustrates the versatility of the yeast cloning system for the production of complex structures.

# 6

# LARGE-SCALE FERMENTATION

The design and operation of contained fermentors are described. The physical features of contained fermentors, the computer system, and guides for environmental monitoring, validation, and typical operations are covered. Cell growth is discussed relative to both genetic control via product regulation as well as exogenous physical or physiological control. Finally, consideration is given to the fermentation and accumulation of product relative to both plasmid stability in fermentors and proteolytic destruction of abnormal proteins (recombinant gene products).

There has been an immense increase in recombinant DNA research since the National Institutes of Health (NIH) first announced guidelines for conducting such research in 1976. Currently, a large variety of homologous and heterologous gene products can be synthesized and stored or excreted by several types of cells. Cells producing recombinant gene products with special medical, agricultural, chemical, or biological properties are all good candidates for large-scale fermentation development. These recombinant gene products may range from more efficiently produced products already synthesized by the cell to molecules not normally produced by these microorganisms. In both cases, it is evident that there are clear logistical advantages associated with isolating the accumulated product from high density, rapidly growing cell populations rather than from the alternatives. Two of these processes, the production of human insulin by combination of recombinant-derived A and B chains and the transformation of recombinant-derived human proinsulin into insulin, have been described in detail.

Two major goals in fermentation process development are to optimize the bioreactor cell concentration and to maximize the product accumulation. Fermentations producing recombinant gene products are similar in many respects to antibiotic-producing fermentations. However, considerable thought and effort have been expended in designing, building, testing, and operating contained fermentation equipment. The purpose of this chapter is to discuss these aspects of contained fermentation technology based on our experience acquired at Eli Lilly while developing the human insulin A and B chain and the human pro insulin fermentations.

## Design and Operation of Contained Fermentors

Since 1976 many of the issues concerning contained fermentor design have changed considerably, though one of the primary concerns, that of personnel safety, has remained unchanged. In 1978, the guidelines for rDNA research restricted fermentation operations to 10L or less in P-3 level containment conditions. The 10L limit was originally imposed arbitrarily by scientists at the Asilomar Conference as a volume that could be conveniently handled in the laboratory. While there was a clear understanding that industrial-level activity was excluded, there was no suggestion that increased volume of a culture of a safe organism would result in any increased risk. We also felt that there was nothing intrinsically more hazardous in volumes exceeding 10L. In fact, most larger operations are carried out in stainless steel vessels, where, with the efficiency of size, fewer operations are necessary. It would seem potentially more hazardous to carry out several hundred glass shake flask fermentations or multiple 10L fermentations than one large fermentor operation in order to produce significant quantities of material. In addition, the product, a chimeric protein consisting of a portion of a lacZ gene peptide or a portion of the trpE gene product and the A or B peptide of insulin or proinsulin was not biologically active.

The revised NIH Guidelines in 1978 included for the first time a short section entitled "Voluntary Registration and Certification". A considerable portion of the pharmaceutical as well as the fledgling genetic engineering industry viewed this as a clear opportunity to begin scaling up laboratory-sized experiments to commercially feasible size. Thus, Lilly and most of the industry have voluntarily complied with the Guidelines and have actively participated in the discussion surrounding each pertinent revision.

Lilly's experience in the fermentation industry dates back to the 1940s and has mostly involved the use of streptomycetes and fungi for the manufacture of antibiotics. Some experience has been gained in several development projects employing *Bacillus* and *E. coli*. In 1968, Lilly successfully produced L-asparaginase from *E. coli* B in large volume for use in experimental cancer chemotherapy.

Although in recent years many technological changes have occurred in the fermentation process, the basic principles for fermentation of rDNA-containing organisms are essentially the same as those used for antibiotic-producing microorganisms since the 1940s. Thus, the various technologies employed in our contained fermentations are not new even though the applications may be novel. The basic equipment consists of a closed stainless steel vessel designed, built, and tested to comply with pressure vessel codes. Such a vessel is normally fitted with an agitator to mix the contents and aid in gas transfer. Compressed air, which has been filtered to remove contaminating organisms, is pumped into the fermentor to provide necessary oxygen for growth. Metabolic heat is removed by a cooling coil or jacket. Modern computing instrumentation monitor and/or control the variables such as fermentor temperature, air flow, pressure, pH, dissolved oxygen, feed rates of various nutrients, and composition of exhaust gases. Operating pressures are modest and are nominally about 5 psig.

Fermentors built for antibiotic production are designed to exclude contaminating organisms. To adapt these vessels for use with recombinant cells, it was necessary to modify them to contain the fermentation organism as well. We modified existing pilot scale antibiotic fermentors to comply with the spirit of existing guidelines, since these guidelines dealt with laboratory containment, not large-scale fermentor containment.

Primary fermentor containment was accomplished by alteration of standard fermentors and operations as follows:

1. All exhaust gases were filtered.
2. Double mechanical agitator seals were installed at the junction of the fermentor and agitator shaft.
3. All inoculations and sampling of a vessel were conducted using a closed, non-aerosol-producing system.
4. The recombinant organisms were chemically or thermally inactivated or killed prior to processing.

In addition, other important aspects of operating contained fermentations included the following:

1. The environment was monitored to verify containment.
2. Detailed procedural protocols were written.
3. Fermentation personnel received extensive training.
4. The system was thoroughly validated.

## Primary Containment Features

With the high volume of air required to support high density cell cultures, there is considerable entrainment of broth with the exhaust air. Most of this entrained liquid is returned to the fermentor by a sterile separating device. The air then passes through a coalescing filter where fine droplets of liquid are removed. The exhaust air is heated in a steam-charged heat exchanger to raise the temperature of the air above the dew point before it passes through a pair of sterilizing-grade air filters arranged in series. The sterile exhaust only then is vented to atmosphere.

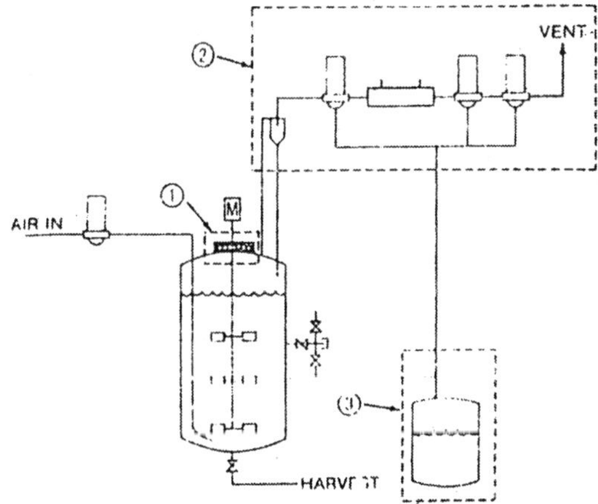


*Fig. 6.1. Primary containment features on fermentors for recombinant gene product production include 1. agitator seal, 2. exhaust air sterilization system, and 3. a condensate collection system.*

The single most likely place for a pressure vessel to leak is at the junction of the agitator shaft with the fermentor vessel. The likelihood of a leak developing at this point is minimized by installing a double mechanical seal at the critical juncture. Either high pressure sterile air or steam/condensate is passed between the seals to ensure containment.

Removing samples from contained vessels is accomplished with a small stainless steel sampling container in which a vacuum is drawn. After attaching the container to the fermentor, the union is sterilized, the sample drawn, and the union finally resterilized. The sampling device is then removed from the tank and taken to a laboratory where the contents are appropriately processed.

For fermentation operations to be carried out in truly contained fashion, the condensate generated from the non-sterile side of the exhaust filters and the broth-containing steam condensate from the sampling device union have to be effectively inactivated. These liquid streams are collected in a condensate collection tank and held until the end of the fermentation. At that time, the system and contents are sterilized and sent to the drain.

Our contained fermentors were built and operated in two distinct generations. Both utilized welded stainless steel vessels and lines. The first generation fermentors were placed within a P-2 containment facility while second generation construction took place in either the fermentation pilot plant or the production fermentor house. Both generations of design included double mechanical seals around the agitator shaft at the junction with the vessel. In operation, the space between the seals is purged with high pressure sterile air or steam/condensate depending on the size of the equipment. First generation equipment was operated under slight negative pressure while later designs permitted operation under positive pressure. Both generations used the extensive exhaust gas filtration system described earlier as well as the condensate collection system. Finally, the cells in first generation fermentors were chemically inactivated while subsequent systems employed chemical inactivation or the thermal inactivation described later in this chapter. Of course, all contained fermentation operations were carried out by specially trained operators using written operating protocols with validated equipment. Environmental monitoring has continued throughout both phases.

**Table 6.1. Elements of contained fermentors.**

| *Item* | *First Generation* | *Second Generation* |
|---|---|---|
| Vessel/lines | All stainless | All stainless |
| Site | Within P-2 facility | None specific |
| Agitator seal | Double mechanical with high pressure air/steam | Double mechanical with high pressure air/steam |
| Back pressure | Negative to atmospheric | Positive to atmospheric |
| Exhaust system | Complex sterilizing filters | Complex sterilizing filters |
| Condensate collection system | Present | Present |
| Cell inactivation | Chemical | Heat or chemical |

## Ancillary Containment Features

Environmental containment monitoring is carried out during each fermentation and includes microbiological sampling of the air from the exhaust system after filtration by using impingers of our own specifications. In addition, sampling of the operator environment is accomplished by using our impingers and a Biotester. The integrity of the exhaust tilter system is determined by DOP testing each filter assembly. After each chemical or thermal inactivation, the fermentation broth is carefully tested to monitor the effectiveness of the inactivation procedures. More than 8000 standard environmental samples have been taken through June, 1983. No isolations of recombinant cells from these samples have been made throughout this period. Fermentations with cells containing the recombinant genes are completely inactivated prior to initiation of further processing. Verification of complete inactivation is carried out in every case prior to disposal of any spent fermentation broth through normal waste handling channels.

**Table 6.2. Environmental containment monitoring.**

| *Sample Source* | *Number Tested* |
|---|---|
| Fermentor Exhaust System | 3192 |
| Operator Environment | |
| Air Samples | 1901 |
| Contact Plates | 2522 |
| Verification of Inactivation | 851 |
| TOTAL | 8466 |

The written protocols that are required to operate fermentation vessels with microorganisms containing rDNA are somewhat complex because of the containment features for inoculation, sampling, etc., but are readily learned by skilled pilot plant and production operators. In addition, these operating protocols also incorporate a number of redundancy features to minimize inadvertent breach of containment. For instance, instructions may be given to set up a critical steam block on the back side of a particular valve and later instructions require verification that the key containment operation has been carried out. Procedures involving crucial elements of containment require the signatures of the operator and his supervisor.

The training of fermentation operators includes a general introduction to various aspects of rDNA technology and containment issues, and highly specific instructions for everyday operation of the facility. The general introduction involves discussions with the biological safety officer, engineering design people, and immediate supervisors concerning various aspects of recombinant DNA, potential risks, containment, NIH guidelines, etc. Detailed specific training on operating protocols designed to meet the specifications of the NIH Guidelines and to avoid loss of containment are carried out by the principal investigator and the operators' supervision.

When the modification of fermentors with these special design features is completed, the entire system is validated. The purpose of validation is to verify that a specific inactivation or sterilization process does what it is supposed to do. To be more specific, when thermal sterilization is employed, validation means measuring the temperatures at multiple points throughout the fermentor and its processing lines. We have found that heat distribution within such complex systems is not as straightforward as it might appear. The findings of these studies periodically have led to changes in fermentor or piping design to correct heat flow problems. Of course, revalidation or the system is completed prior to initiating contained operations.

Validation of the broth thermal inactivation process was completed in two distinct phases. The first operation was to quantitate the death rate or D-values (lethality) for *E. coli* K-12, under defined conditions in the laboratory. Two such determinations are shown in Figures 6.2 and 6.3 for 56°C and 60°C, respectively. Such relatively low temperatures were used in these heat kill studies so that individual

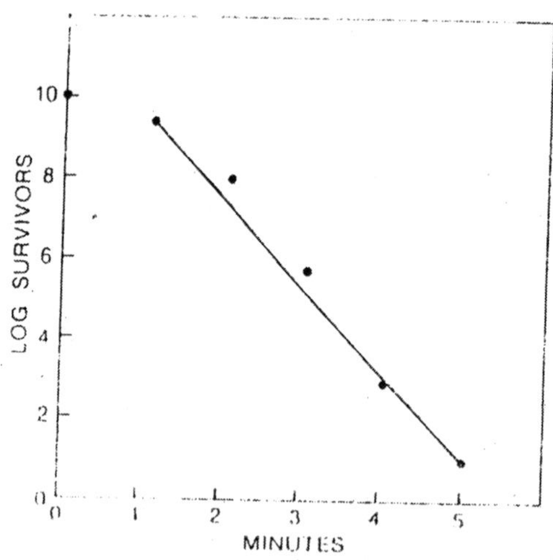


*Fig. 6.2. Log of the number of E. coli, K12, strain RV308, survivors versus the time the cells were held at 56°C. The D-value is the inverse of the scope.*

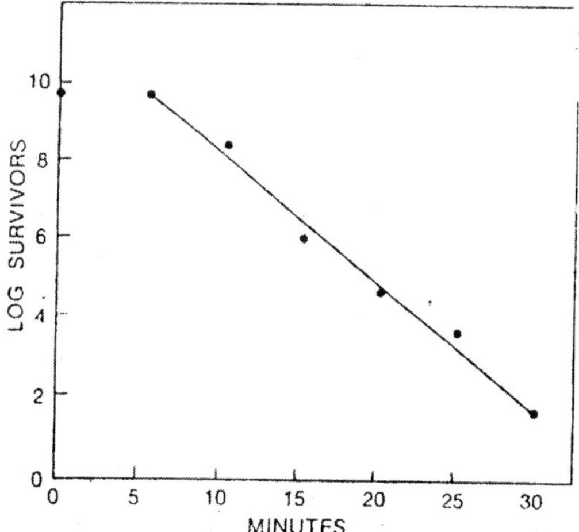


*Fig. 6.3. Log of the number of E. coli, K12, strain RV308, survivors versus the time the cells were held at 60°C. The D-value is the inverse of the scope.*

samples could be accurately taken. At 56°C, eight logs of kill occurred over 30 minutes, whereas at 60°C the same degree of cell inactivation occurred within five minutes. The D-values determined from these and other studies are defined as the inverse of the slope at each temperature or the time required to reduce the survivor population by one log. The z-value for this process can be graphically represented by plotting D-values versus temperaure. The z-value is the inverse of the slope and represents the number of degrees centigrade required to change the D-values by a factor of 10. These values, that is, the D- and z-values are used to calculate the log reduction in viable organisms per minute at any temperature.

The second part of the validation process was to apply the laboratory defined conditions to the fermentors. Under the conditions we established for heat inactivation, calculations predicted that the system delivered well over the required log reduction in viable cells, which included a six log safety factor. Such large margins of safety are not necessary but can be utilized if product stability is not a factor. In addition to gathering temperature measurements during this second validation phase, we were also concerned with the overall success and reproducibility of the inactivating process. Thus, the validation team evaluated the inactivation strategy as well as observed the interaction between the fermentation operator, the equipment, the computer, and the written protocol. The team was satisfied that these various components complemented one another such that the process was reliable.

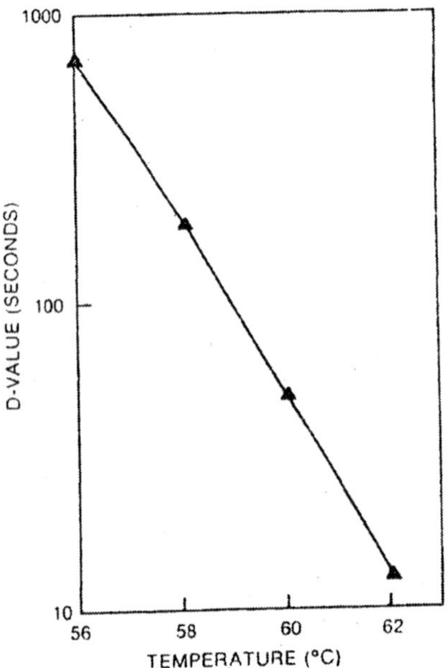


*Fig. 6.4. Log of the D-value determined from Fig 6.2, 6.3 and other data as a function of the temperature. Z-value is the inverse of the scope.*

## COMPUTER SYSTEM

Modern antibiotic or recombinant fermentation process control requires the use of computers for adequate control of process variables. The fermentation process control computer system at Eli Lilly has been recently described and consists of a hierarchy of computers arranged by function. The Level 1 system consists of several 32K word mini-computers, one of which may drive five to ten contained fermentors. It handles setpoint control, collection of data, and alarming functions. Limited data storage is available at this level, but most data is communicated to the next level.

The Level II system is a 192K word mini-computer, and normally receives data from several Level I computers. Its primary function is data management. To accomplish that task, there are several peripherals which include: character printers, graphic CRTs, plotters, line printers, magnetic tape cartridge recorders, modems for remote terminals, large storage discs, and nine track tape. While most data handling is done at this level, a communication link is available to the corporate computer. The Level III corporate computer can receive some selected data and is available for large model building. In addition, several statistical packages are available on the Level III. In addition to the normal control functions routinely carried out, the process control computer can maintain control over crucial containment elements. Sensors placed at appropriate positions in the equipment relay this containment information to the computer. Values seen as sufficiently outside the setpoint control value require immediate action by the computer to prevent a breach of containment. The computer either readjusts a set point or shuts the fermentation down. In either case, it alarms and prints a message concerning the particular problem. There is no substitute for the constant level of surveillance supplied by the fermentation computer.

## Cell Growth and Product Regulation

The major goals in fermentation development are to optimize cell mass and maximize product accumulation. Occasionally, these goals, which usually coincide, may conflict with one another. Such problems arise in host/vector combinations where the expression of the desired recombinant gene product is driven by a strong constitutive promoter/ operator region. The unregulated accumulation of gene product may begin to seriously impair physiological growth functions. For example, when cells containing a regulatable recombinant gene (chimeric human insulin A chain) are grown under totally derepressed (unregulated)

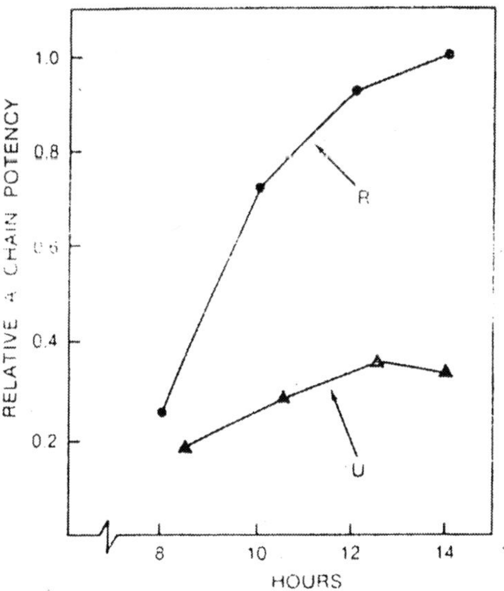


*Fig. 6.5. A chain accumulation in recombinant gene-regulated (R) and -unregulated (U) fermentations versus the termination hour.*

conditions, the volumetric product accumulation is fractional to that produced during a well-regulated fermentation cycle (i.e., the recombinant gene is repressed during early cell growth and derepressed at eight hrs to allow synthesis of product). The difference in cell mass accumulation, does not account for the entire difference in volumetric activity. Thus, there is considerable improvement in the specific activity as well. It is desirable, then, that expression of a recombinant gene be placed under some reasonable form of control.

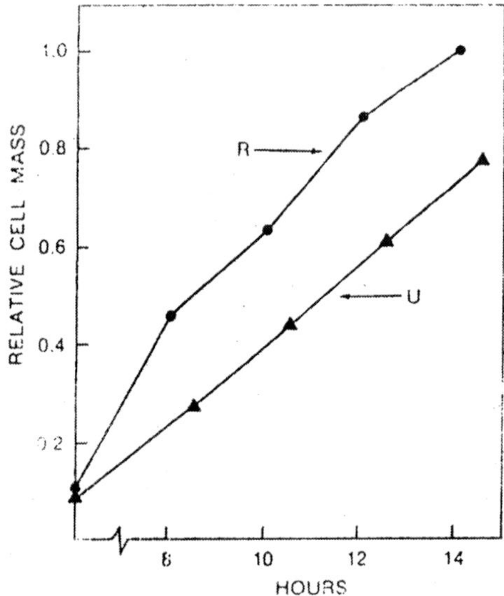


*Fig. 6.6. Cell mass accumulation in recombinant gene-regulated (R) and -unregulated (U) fermentations versus the fermentation hour.*

This control might be inducible as in the case of the control elements for the *lac* operon or the penicillinase gene, or end product regulated as in the case of the *trp* operon promoter/operator. The necessary control may also be exercised by amplification of a recombinant gene through temperature-regulated "runaway" replication of the plasmid.

## Control of Cell Growth

Limited cell growth in the fermentor using partially auxotrophic strains of *E. coli* K-12 (all K-12 strains require thiamine) can be carried out in either a variety of complex media such as L broth or others or in defined media such as M9 medium or other defined media such as those used to determine maximal growth rates in *E. coli*.

The limiting factor in most fermentation operations is that the oxygen transfer capability of a fermentor is ultimately exceeded by the demand for oxygen by the culture. To control that demand, one or more of the growth substrates are limited in either fed-batch culture or in continuous culture.

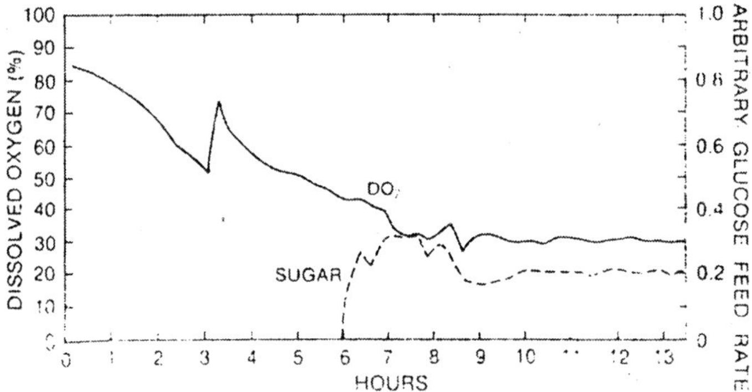


*Fig. 6.7. Control of dissolved oxygen at approximately 30% by glucose feed versus the fermentation time.*

In a fed-batch experiment (with human insulin A chain) dissolved oxygen, an easily measured variable associated with the oxygen transfer capability of a vessel, was maintained at 30 percent by means of a glucose feed regulated by closed loop control maintained by the computer. As in the case of antibiotic fermentations using this sort of strategy, normal cell mass and potency were obtained. Such a feeding strategy is, in fact, practically mandatory in systems where catabolite repression influences the regulation. In some cases, however, the

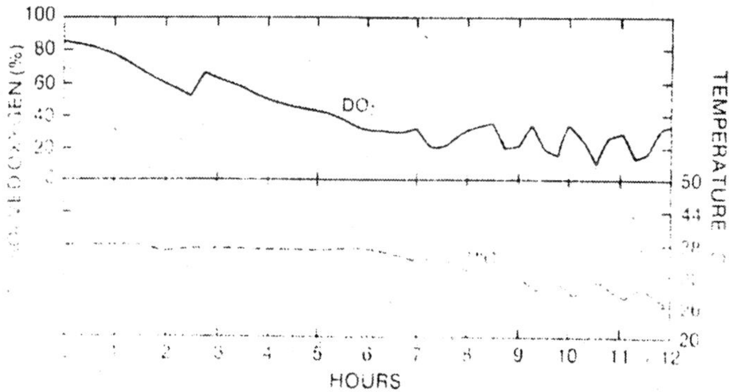


*Fig. 6.8. Control of dissolved oxygen at approximately 30% by temperature control of oxygen demand versus the fermentation time.*

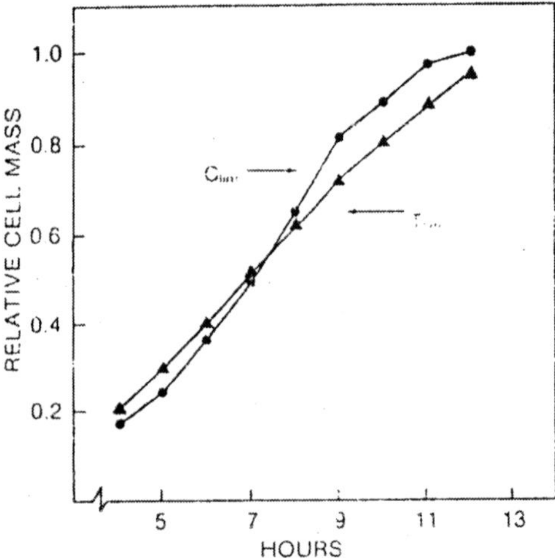


*Fig. 6.9. Cell mass accumulation during the carbon ($C_{lim}$) or temperature limit ($T_{lim}$) growth versus the fermentation time.*

nutrient limitation imposed to control growth is sufficient to induce intracellular proteases.

More recently, controlled cultivation of microorganisms in which no nutritional limitation is imposed has been described. The demand for nutrients was limited by controlling the growth temperature of the culture. Growth at reduced temperatures has been shown to increase

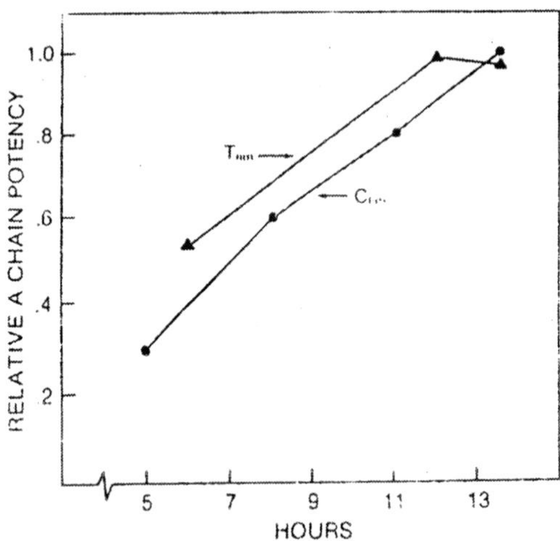


*Fig. 6.10. Accumulation of A chain during carbon ($C_{lim}$) or temperature limited ($T_{lim}$) growth versus the fermentation time.*

the cell yield on oxygen and maximal exponential cell yields. This approach has been reported to yield nearly 50 gm dry cells/L in pilot scale fermentations with *E. coli* W. We have never achieved such dramatic cell mass accumulations in recombinant gene-containing fermentations, since product accumulation eventually affects growth. We were, however, able to control dissolved oxygen concentrations during the cell growth phase by temperature. Temperature control and carbon source growth control strategies have been compared and found to be practically indistinguishable, both in the literature and in our own experience.

## Product Biosynthesis and Accumulation

A recombinant gene product is synthesized in response to the recombinant gene originally cloned into the appropriate vector. The level of product formed is a function of several elements including:

1. The relative strength of the promoter
2. The stability of the transcriptional and translational elements
3. The stability of the plasmid within the host
4. The level of intracellular protease involvement

Presumably promoter strength, and transcriptional and translational problems are addressed during the process of cloning the gene and will not be discussed here.

The rate at which product is formed is partially a function of plasmid stability. The plasmid-minus cell is, of course, incapable of synthesizing any recombinant product, and has a significant growth rate advantage over the plasmid containing cell. The plasmid-minus cell has an even greater advantage over plasmid-containing cells in continuous culture where the number of cell divisions is much greater than fed-batch culture. However, there is some evidence that adaptive changes may occur in plasmid-containing organisms which increase their growth rate and result in a delayed loss of the plasmid from the population. To maintain plasmid-bearing cells, it is necessary to utilize a relatively high copy number plasmid, add selective pressure to the fermentation medium, utilize some plasmid stabilizing feature of the vector itself, or alter growth conditions to maintain the recombinant gene in 100 percent of the population. We have had good success using tetracycline selection to maintain the proinsulin plasmid in virtually all cells during large-scale fermentations. For example, samples which were taken during an experimental fermentation were examined for plasmid retention by spotting isolated colonies on selective and non-

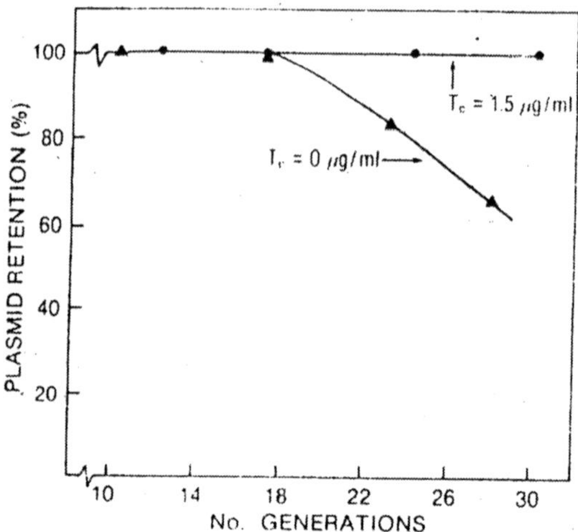


*Fig. 6.11. Stability of proinsulin vector with and without tetracycline selection as a function of the number of generation under those conditions.*

selective media. All colonies originally grown in the presence of tetracycline still contained the plasmid after 30 generations, while control colonies grown in the absence of the selective pressure were segregating plasmid-minus cells after approximately 18 generations.

The level of product accumulation during fermentation may also be affected by induced intracellular proteases. It has been shown that the same family of peptidases that hydrolyze peptides from substrates

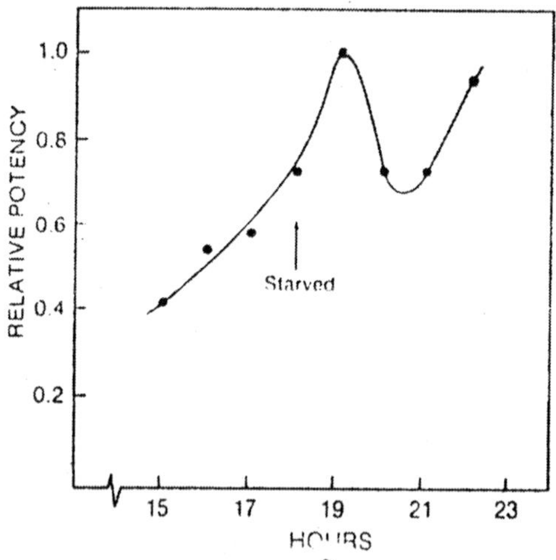


*Fig. 6.12. Product loss as a function of starvation for glucose.*

during normal growth also are responsible for hydrolysis of peptides during periods of starvation for carbon. The same is true of cells starved for nitrogen or minerals. Sufficient care should be taken during the design of the original medium and control strategy such that starvation does not occur during growth. Conditions imposed to control oxygen demand must also be carefully established lest starvation conditions be initiated. During one experiment in which glucose was being fed to control dissolved oxygen, starvation was deliberately induced for a period of 15 minutes. The soluble product being studied began disappearing after a short lag period. After control was reestablished, net positive product accumulation was again seen after approximately two hours. Little deviation in product accumulation is seen, however, when an insoluble recombinant gene product accumulates. The identity of the insoluble recombinant product with the inclusion bodies that form has been established. Pulse/chase experiments done with B chain chimeric protein-containing cells reveal little, if any, turnover of the insoluble product during the fermentation. Thus, deposition of the recombinant product as an insoluble inclusion body appears to confer stability.

# 7

# High Fructose Corn Syrup

From its relatively modest beginning in the late 1960s, the manufacture of high fructose corn syrup has registered an impressive growth. *High fructose corn syrup* (HFCS) has become an established product line for the corn wet milling industry with almost every major corn processor in the United States having one or more HFCS plants in operation. In 1982, high fructose corn syrup production amounted to 8.4 billion pounds with a value well in excess of one billion dollars.

## Historical Perspective

The evolution and technology of sweeteners is an interesting history that encompasses some of the most significant events in chemistry and biotechnology. In some respects, the developments of high fructose corn syrup—a mixture of dextrose and fructose—completes a circle back to wild honey, the oldest sweetener known to man. This circle also includes sucrose and the wide variety of corn syrups manufactured from corn starch.

It is believed that cane sugar dates back at least 8,000 years to the South Pacific. The sweetness of cane sugar approximates that of wild honey. It wasn't until 1744 that a German chemist found that the sugar isolated from sugar beets was identical to the sugar from sugar cane. Sucrose, manufactured from cane or sugar beets, was very much more abundant than honey and as such, became the sweetener of commerce. This position was unchallenged until the development of high fructose corn syrup.

The development of high fructose corn syrup can be traced back to 1811 when the Russian chemist Kirchoff reported his discovery that starch yielded a sweet substance when heated with acid. In 1815, de

Saussure determined that the reaction of Kirchoff on starch was hydrolysis. He found that the end product of the hydrolytic reaction was grape sugar or dextrose.

These experiments established the basis for the commercial production of starch syrups and crude starch sugars. As the demand for starch derived syrups in. creased, corn emerged as the preferred source of starch because it was low in cost, plentiful, and easily stored. The result was the beginning of the present day corn wet milling industry. Significantly, this development opened up the corn growing temperate agricultural zones of the world to the production of raw materials for sweeteners with profound and continued impact on national economics. Heretofore, sweetener production had been largely limited to the tropical and semi-tropical agricultural zones where sugar cane is grown.

## Development of Key Enzymes

The discovery, isolation, and application of various carbohydrase enzymes resulted in the development of many new corn syrups. These enzymes, principally bacterial alpha amylase, fungal glucoamylase, and fungal alpha amylase could be produced cheaply by fermentation processes. These enzymes could be used singly, in sequence, or in combinations on starch to produce a broad range of syrup compositions with different properties. These products were called corn sweeteners.

The commercial development of glucose isomerase which converts glucose to its sweeter isomer fructose was a major milestone in the corn sweetener industry. A historical perspective of this exciting development is presented by Casey.

The enzymatic transformation of glucose to fructose was first introduced to corn sweetener production in 1967. The first HFCS produced contained 15 percent fructose. The manufacturing process, known as isomerization, originally involved the direct addition of an isomerase enzyme to a dextrose substrate in a batch process. Further process improvements resulted in the HFCS product containing 42 percent fructose.

In 1972, batch production was replaced by a continuous process utilizing isomerase enzyme immobilized on an insoluble carrier. This continuous process for the production of high fructose corn syrup involved the first large scale use of an immobilized enzyme system in the world. The continuous process resulted in significant reduction in production costs. The design and operation of continuous immobilized isomerase reactors for HFCS has been described. A description of the products of modern process technology will facilitate the detailed discussion of essential features of saccharification, isomerization, and fractionation in the process.

## High Fructose Corn Syrup Products

Three high fructose corn syrup products differentiated by fructose content are commercially available. These products contain 42, 55, or 90 percent fructose.

**Table 7.1. Carbohydrate composition of high fructose Corn syrup produced.**

| | *Typical Analysis* | | |
|---|---|---|---|
| *Carbohydrate* | *42 HFCS* | *55 EFCS* | *90 VEFCS* |
| Fructose | 42 | 55 | 90 |
| Dextrose | 52 | 42 | 7 |
| Monosaccharides | 94 | 97 | 97 |
| Higher Saccharides | 6 | 3 | 3 |

**Table 7.2. HFCS producers**

| *Producer* | *Plant Location* | *Plant Size (bu/day)* | *HFCS Capacity (lbs. C.B./yr.)* |
|---|---|---|---|
| Archer Daniels | Cedar Rapids, IA | 220,000 | 1,264,000 |
| Midland | Decatur, IL | 200,000 | 1,714,000 |
| American Maize | | | |
| Products | Decatur, AL | 35,000 | 500,000 |
| Amstar | Dimmitt, TX | 30,000 | 336,000 |
| Cargill | Dayton, OH | 150,000 | 696,000 |
| | Memphis, TN | 90,000 | 696,000 |
| Corn Products | Argo, IL | 145,000 | 551,000 |
| Company | Stockton, CA | 32,000 | 352,000 |
| | Winston-Salem, NC | 32,000 | 351,000 |
| Clinton Corn | Clinton, IA | 110,000 | 948,000 |
| Processing | Montezuma, NY | 35,000 | 423,000 |
| Hubinger | Keokuk, IA | 83,000 | 632,000 |
| A.E. Staley | Decatur, IL | 150,000 | 600,000 |
| | Morrisville, PA | 85,000 | 700,000 |
| | Lafayette, IN | 155,000 | 1,300,000 |
| | Loudon, TN | 70,000 | 600,000 |
| | | TOTAL = | 11,663,000 |

High fructose corn syrups containing 42 percent fructose can be used in most food products that make use of a liquid sweetener. The development of second generation syrups with higher fructose levels greatly extended the use of HFCS in many applications. This is particularly true with respect to the use of HFCS with 55 percent fructose as a replacement for sucrose in soft drinks.

Fructose corn syrup containing 90 percent fructose has found wide application in many new and specialty food products. It is an ideal sweetener for low calorie foods such as jams and jellies.

Today, there are eight companies manufacturing high fructose corn syrup. These companies, together with their location, plant size, and HFCS production capacities.

## Modern Process Technology

The manufacturing process may be divided into about 18 separate steps in five major phases of operation, each of which has a major objective. These phases include dextrose production, primary refining and chemical treatment to produce dextrose feedstock, isomerization of dextrose feedstock to produce 42 percent fructose, secondary refining of 42 percent fructose, and fractionation of 42 percent fructose for the production of 55 percent fructose.

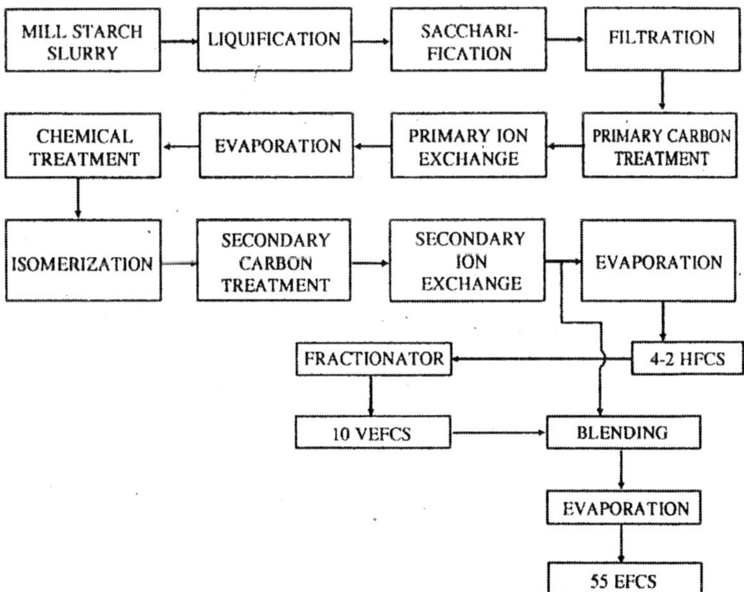


*Fig. 7.1. High fructose corn syrup manufacture simplified process schematic.*

The heart of the entire process is the enzymatic isomerization step utilizing immobilized glucose isomerase.

The economics and success of the isomerization reaction depends upon the production and delivery of high quality dextrose feedstock to the isomerization reactors.

## Manufacture of Dextrose from Starch

The manufacture of dextrose feedstock from starch is a multi stage process involving thermostable alpha amylase and glucoamylase in successive enzymatic steps. These enzymes catalyze the hydrolysis of the starch polymer to dextrose monomer.

In the first step, mill starch slurry is gelatinized by cooking at high temperature. The gelatinized starch is then liquified and dextrinized by thermostable alpha amylase in a continuous two stage reaction. The product of this reaction is a soluble dextrin hydrolyzate with a D.E. (dextrose equivalent) of 10-15, suitable for the following saccharification step.

During the liquifaction-dextrinization reaction, the major process variables of starch quality and concentration, alpha amylase dose, temperature, pH, starch flow, and time must be maintained within relatively narrow limits to insure that the starch is properly cooked and dextrinized. If anyone of these variables is significantly out of process specifications, the process liquor may contain undercooked and unconverted starch.

**Table 7.3. Liquifaction-dextrinization reaction conditions.**

| | | |
|---|---|---|
| Starch Slurry | | 30-35% w/w d.b. |
| Total Protein | | ≤0.3% |
| Soluble Protein | | ≤0.03% |
| pH | | -6.0-6.5 |
| Thermostable Alpha Amylase | | -0.05-0.10% d.b. starch |
| First Stage | Temperature | -104-107°C |
| Liquifaction | Hold | -5-8 min |
| | D.E. | -0.5-1.5 |
| Second Stage | Temperature | -94 -97° |
| Liquifaction | Hold | -90-120 min. |
| | D.E | -10-15 |

This leads to poor conversion and filtration problems, or the production of anomalous saccharides such as maltulose which can lead to flavor problems in the finished high fructose corn syrup.

The total and soluble protein levels in mill starch slurry should be maintained below about 0.3 percent total protein and 0.03 percent soluble protein. This is necessary to minimize colour formation as a result of the Maillard reaction between amino acids and sugars under conditions of high temperature and pH. If the mill starch slurry is high in protein, the syrup produced has a greater potential to develop colour exceeding the capacity of the carbon and ion-exchange refining systems.

Following liquifaction-dextrinization, the pH and temperature of the 10-15 D.E. hydrolyzate is adjusted for saccharification. During saccharification, the hydrolyzate is further hydrolyzed to dextrose by the action of glucoamylase enzyme. Although saccharification can be carried out as a batch reaction, a continuous saccharification is carried out in most modern plants. In the continuous saccharification reaction, glucoamylase is added to the 10-15 D.E. hydrolyzate following pH and temperature adjustment. The saccharifying liquor is then pumped through a number of large reactors in series.

**Table 7.4. Saccharification reaction conditions.**

| | |
|---|---|
| Dry Substance | 30-35% |
| pH | 4.0-4.4 |
| Temperature | 60-62°C |
| Glucoamylase | 1 liter/ton d.s. starch |
| Reaction Time | 65-75 hours |

The total time for saccharification is particularly sensitive to the amount of glucoamylase. Commonly, a reaction time of 65-75 hours is used to obtain a hydrolyzate containing 94-96 percent dextrose.

During saccharification, the important variables monitored are the initial D.E., pH, temperature, dry substance, glucoamylase dose, the time course of conversion, and the presence of unconverted starch in the liquor during this reaction.

**Table 7.5. Dextrose saccharification product typical analysis.**

| *Component* | *Percent, d.b.* |
|---|---|
| Dextrose | 94-96% |
| Maltose | 2-3% |
| Maltotriose | 0.3-0.5% |
| Higher Saccharides | 1-2% |
| Dry Substance | 35-37% |

These variables must be controlled within relatively narrow limits to insure that a minimum 94 percent dextrose level is reached within relatively narrow time limits. This minimum level of 94 percent dextrose is necessary to meet the specifications for the carbohydrate composition of 42 percent HFCS in which the higher saccharides level cannot exceed 6 percent. The high dextrose hydrolyzate is then refined to produce dextrose feedstock for isomerization to 42 percent high fructose corn syrup.

## Preparation of Dextrose Feedstock for Isomerization

Preparation of very high quality dextrose feedstock for isomerization is first necessary because of the very low colour and ash specification for high fructose corn syrup. A high purity feedstock is also required for efficient utilization of the immobilized isomerase enzyme column.

Immobilized isomerase enzyme columns are used continuously over a period of several months. During this period very large volumes of dextrose feedstock pass through the columns. Extremely low levels of impurities such as ash, metal ions, or protein in the feedstock can accumulate and lead to decreased productivity of the enzyme. For these reasons, dextrose feedstock is refined to a colour of 0.1 (CRA × 100) and a conductivity of 5-10 $\mu$mhos.

In the refining process, the high dextrose conversion liquor is filtered on rotary precoat filters to remove "mud" or coagulated protein and oil. This filtered liquor is then passed through a series of check filters and polish filters to remove traces of particulate matter.

Colour in the hydrolyzate is first removed in a series of granulated carbon columns. These columns are renewed by the removal of spent carbon and the addition of regenerated carbon on a regular basis.

Carbon treated liquor is filtered again and passed through an ion-exchange system where it is deionized by ion-exchange resin. This resin is usually arranged in a dual pass cation-anion-cation-anion system. Lon-exchange is designed to remove all heavy metals and ash that might be detrimental to the immobilized isomerase enzymes. Ion-exchange resin also removes significant amounts of colour bodies that are not removed by the carbon columns.

The carbon-treated and deionized high dextrose liquor is evaporated to the proper solids level for isomerization. In addition, the feedstock is chemically treated by the addition of magnesium ions which not only activate the immobilized isomerase but also competively inhibit the action of any residual calcium ions which are potent inhibitors of isomerase.

## Isomerization Process

The design and operation of a commercial immobilized glucose isomerase reactor system has been described. Careful integration of design and operation with respect to reaction conditions are necessary to minimize production fluctuations with respect to capacity and conversion level.

**Table 7.6. Isomerization feedstock and reaction conditions.**

| | |
|---|---|
| Dry Substance | 40-45% |
| Dextrose | 94-96% |
| Higher Saccharides | 4-6% |
| pH | 7.5-8.2 at 25°C |
| pH Drop | 0.2-0.4 |
| Temperature | 55-65°C |
| Activator | 0.004 M $Mg^{++}$ |
| Residence Time | 0.5-4 hr. |

The conversion of glucose to fructose is a reversible reaction with an equilibrium constant of about 1.0 at 60°. Thus, one would expect to obtain a fructose level of about 47-48 percent at equilibrium, starting from a feedstock containing 94-96 percent dextrose. However, the reaction rate near the equilibrium value is so slow that it is prudent to terminate the reaction at a conversion level of 42 percent fructose to achieve practical reaction residence times.

In a given isocolumn (immobilized isomerase column), the rate of conversion of dextrose to fructose is proportional to the enzyme activity of the immobilized isomerase. This activity decays in a regular manner as a function of time. When the column is new and the activity is high, the flow of feedstock through the column is relatively high since a shorter residence time is required to achieve the 42 percent fructose level. The flow through the column must be reduced proportionately to give a longer residence time compensating for the lowered activity in order to achieve a constant conversion level as the column ages.

Since the activity of the bound enzyme decreases continuously, it is necessary to have a number of isocolumns to minimize production fluctuations with respect to capacity and conversion level. Both series and parallel operation of the isocolumn are possible. In practice, the parallel operation of at least six isocolumns offers the greatest operational flexibility. In this arrangement, each isocolumn can be operated essentially independent of the others. Each isocolumn can be brought into and taken out of service very readily.

It is necessary to operate the isocolumn battery to obtain a finished product of uniform quality and constant average fructose content of 42 percent. The variation in total flow of the isocolumns must be maintained within relatively narrow limits because of the requirements of evaporation and other finishing operations.

In practice, flow cannot be precisely controlled at all times to obtain 42 percent fructose, but this can easily be achieved on an averaged basis.

A number of important process control strategies can be used to achieve an approximate steady state operation. These include modest variations in operating temperature, conversion level, and feedstock pH control. Final uniformity of product quality and fructose level is usually accomplished by automatic back-blending operations controlled by an in-line polarimeter.

The effect of temperature on overall productivity can be dramatic. A feed temperature of 60°C is considered as an optimum. Higher temperatures result in faster flow rates but also cause accelerated decay rates. Feed temperature as low as 55°C can be used to extend enzyme life at the expense of slower flow rates. Some risk of microbial contamination does exist at the lower operation temperature.

One of the most critical operating variables is the internal isocolumn pH. The operating pH is usually a compromise between the pH of maximum activity, typically around pH 8, and the pH of maximum stability, typically pH 7.0-7.5. This is complicated by the fact that dextrose feedstock is not stable to pH at temperature around 60°C. Some decomposition to produce acidic by-products results in a pH drop across the isocolumn during operation.

Since it is difficult to control pH directly in the column, pH adjustment is made to feedstock to maintain a constant pH in the column effluent.

The operational stability of the immobilized isomerase system can be characterized by the half life, or the time span during which the enzyme activity is reduced by one-half. Typically, a single isocolumn is operated for a period of at least two half-lives, after which the enzyme is discarded and replaced with a fresh batch. The operating life of a column is determined by such consideration as the number of isocolumns in the battery, the average decay rate of the individual columns, the maximum allowable variance in individual isocolumn flow, minimum total flow, and total required production capacity. Enzyme half-life of 70-120 days are common resulting in column replacement of 1.5-2 times per year.

Isomerization enzyme cost is a significant fraction of total operating cost. The enzyme can be purchased commercially or manufactured by the user. In the former case, most enzyme is supplied on a performance guarantee with minimum stated activity. The quality of immobilized isomerase has improved with respect to longer half-life and higher initial activity with a net of higher productivity. In general, isomerization costs have decreased significantly from the original cost of 50 to 70 cents per cwt. HFCS dry basis.

The overall process economics is improved by every percent increase in enzyme productivity.

**Table 7.7. Principal producers of immobilized isomerase.**

| *Company* | *Trade Name of Product* |
|---|---|
| Gist - Brocades | Maxazyme GI Immob. |
| Miles Laboratories. Inc. | Takasweet |
| Novo Industries | Sweetzyme Q |
| UOP Process Division | Ketomax GI-100 |

Most of these preparations consist of fixed whole cells containing glucose isomerase activity. The products of NOVO and UOP are believed to be manufactured from isolated and purified glucose isomerase.

An objective comparison of the performance of different immobilized enzymes is essential from a user's point of view. Since the enzyme supplier's data are generally provided on different bases, laboratory and pilot plant evaluations are generally required to compare competing systems. The biochemical, mechanical, and hydraulic characteristics of importance in evaluating the suitability of an immobilized enzyme for a particular reactor design have been reviewed.

## Secondary Refining of Isomerized Feedstock

Following isomerization, the manufacturing process involves the secondary or polish refining of the 42 percent HFCS product. Some additional colour is picked up during the chemical treatment and isomerization when the feedstock is held at higher pH and temperature for a period of time. The product also contains some additional ash from the chemicals added for isomerization.

This colour and ash is removed by the secondary carbon and ion-exchange systems. The refined 42 percent HFCS is then evaporated to 71 percent solids for shipment.

## Manufacture of Enriched Fructose Products

The HFCS product from the isomerization reaction typically contains 42 percent fructose, 52 percent unconverted dextrose, and about 6 percent of oligosaccharides. For reasons previously discussed, this product represents the practical maximum level of fructose attainable. In order to obtain products with higher levels of fructose, it is necessary to selectively concentrate the fructose. Many common separation techniques are not applicable for this purpose, since they do not readily discriminate between the two isomers of essentially the same molecular size. However, fructose preferentially forms a complex with different cations such as calcium. This difference has been exploited to develop commercial processes by taking advantage of this property of fructose and combining it with more established separation technologies such as ion-exchange.

There are basically two different commercial processes available today for the large scale purification of fructose. In both instances, resins in the preferred cationic-form are used in packed bed systems. One process employs an inorganic resin leading to a selective molecular absorption of fructose.

Chromotographic fractionation using organic resins is the basis for the second commercial separation process. When an aqueous solution of dextrose and fructose such as 42 percent HFCS is fed to a fractionating column, fructose is selectively held by the resin to a greater degree than dextrose. Deionized and deoxygenated water is used as the eluant. Typically, the separation is achieved in a column packed with a bed of low cross-linked fine mesh polystyrene sulfonate cation-exchange resin using calcium as the preferred salt form. The enriched fructose containing approximately 90 percent fructose is referred to as *very enriched fructose corn syrup* (VEFCS). This VEFCS fraction can be blended back with the 42 HFCS feed material to obtain products having fructose content between 42 percent and 90 percent. The most typical of these products is 55 percent enriched fructose corn syrup, called 55 EFCS.

The treatment of other raffinate streams in the fractionation process is an important consideration. In general, the dextrose rich raffinate stream is recycled to the dextrose feed of the isocolumn system for further conversion to 42 HFCS. The raffinate stream containing dextrose with fructose levels higher than the feed level is generally recycled through the fractionator to maintain high solids level and reduce water usage. The raffinate stream rich in oligosaccharides is recycled back to the saccharification system.

Since water is used as the elution medium, it has a great impact on the overall evaporation load on the system. Very low solids concentrations contribute to the risk of microbiological problems within the system. Thus, the most important design parameter dictating overall process economics is the maximization of solids yield at acceptable purity while reducing the dilution effect of the eluant rinse to a minimum. The efficiency of feed and water usage must be maximized for optimal yield. The yield is important to reduce the cost of reisomerization of each kilogram of solids as well as for other obvious reasons.

Procedures available for achieving these goals include recycling techniques, higher equalization of the resin phase with proper redistribution in a packed column, and the addition of multiple entry/exit points in the column. These approaches to increase the purity and the yield are sometimes referred to as feed enrichment. A small apparent increase in the purity of feed to the column, i.e., higher fructose levels, results in a much larger gain in production through increased yield at a given product purity. In practice, this translates into maximization of the ratio of sugar volume fed per volume of resin per cycle, minimization of the ratio of water volume required per volume of resin per cycle, and provision for careful fluid distribution to the columns.

## Marketing and Economic Considerations

Sucrose has long been the dominant caloric sweetener in the United States. Since 1972, it has been losing ground to HFCS. In 1982, the production of 42 percent and 55 percent fructose corn syrups amounted to over 5 billion pounds dry to account for approximately 28 percent of the total U.S. sweetener market.

The remarkable growth of the HFCS industry in the last 15 years was dramatically triggered by high sugar prices in the early 1970s. In 1971, the wholesale price of sugar (Northeast) was 11.5 cents per pound. By 1976, the price rose to 19.2 cents per pound and in 1980, a record price of 41.0 cents per pound was reached. In comparison, in 1976 high fructose corn syrup (Illinois) sold for 14.0 cents per pound or 73 percent of sugar cost. In 1980, during the record rise in sugar prices, HFCS was selling for 23.6 cents per pound or only 58 percent of the sugar cost. In 1981, sugar dropped about 25 percent to 30.6 cents per pound; HFCS prices dropped only about 9 percent to 21.5 cents per pound or 70 percent of the sugar cost.

The price differential between HFCS and sugar offered strong incentive for the substitution of HFCS for sugar in soft drinks, canned fruits, ice cream, and certain bakery products.

As a result, sugar usage by industrial users has been declining relative to total sugar and particularly relative to total sweeteners because of the inroads of high fructose corn syrup.

This decline in sugar usage is dramatically illustrated by the beverage industry, the largest industrial users of sugar. In 1978, annual sugar usage peaked at 5.12 billion pounds. By 1981, this usage had declined 28.5 percent to 3.66 billion pounds. The total industrial sugar usage declined about 18 percent from 13.9 billion pounds in 1989 to 11.4 billion pounds in 1981. The trend of HFCS substitution for sugar in soft drinks, canned fuits, ice cream, and certain bakery items is expected to slow down as HFCS saturates these markets by 1985.

Caloric sweetener consumption in the United States reached a peak of 129.3 pounds per capita in 1977, after growing at a rate of about one pound per person per year since 1960. By 1982, this had declined to about 122.4 pounds per person. An annual consumption of about 130 pounds per capita has been considered a safe plateau. Future growth in total caloric sweetener consumption is expected to occur only in direct proportion to population growth.

It is expected that sugar will continue to be replaced by HFCS particularly in the processed foods sector. Consequently, the sale of HFCS is projected to grow, but at a decreasing rate.

The beverage industry, particularly soft drinks, offers the greatest growth potential. If the major cola companies were to increase usage, much excess capacity would be used. Currently about 50-60 percent of the HFCS produced is used in the soft drink industry, 14 percent in baking, and seven percent in canning. A large industrial user of 42 or 55 percent HFCS is unlikely to switch back to sucrose as long as an attractive price differential exists and product quality is maintained.

# 8

# FERMENTATION REACTIONS

## METABOLIC GROUPS AND PATHWAYS

### Two- and Three-phase Models

Anaerobic digestion consists of a complex series of reactions, which are catalyzed by a consortium of bacteria and accomplish the conversion of organic compounds to the terminal products methane and carbon dioxide. The sum of these reactions is a fermentation which converts a wide array of substrate materials having carbons at various oxidation/reduction states to one-carbon molecules in the most oxidized ($CO_2$) and most reduced ($CH_4$) states. The latter arrangement is thermodynamically the most stable state (at atmospheric pressure), and these molecules are not further degraded in ecosystems where light and inorganic electron acceptors such as nitrate or oxygen are absent.

The traditional model of anaerobic digestion divides the reactions into two phases, or stages. The first, acidogenesis, rearranges (ferments) the organic matter to carbon dioxide, hydrogen, and volatile organic acids, including acetate, propionate, and butyrate. The methanogenic phase converts these extracellular intermediates to the terminal products, methane and carbon dioxide.

This division into two phases has for many years been a valuable model for anaerobic digestion, because it emphasizes important distinctions between organisms of each phase. For instance, the acidogenic bacteria are generally fast growing and resistant to inhibition, in contrast to the slow growing and fastidious organisms of the methanogenic phase. However the model obscures important interactions between the bacteria of the acidogenic phase and those of the methanogenic phase. Recent advances in the understanding of these

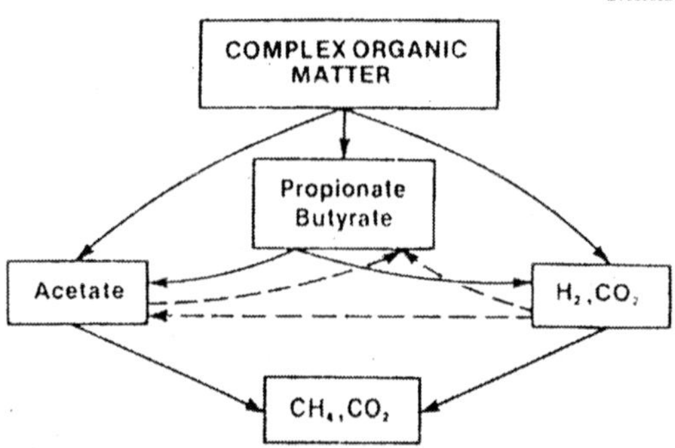


*Fig. 8.1. Three-stage model for complete anaerobic digestion of organic matter to methane and carbon dioxide. Dotted lines represent "black reactions" in which hydrogen is used in making acetate and higher volatile organic acids.*

interactions has led to a more complete model, having three stages instead of two. The traditional model groups together in the methanogenic phase all bacteria responsible for the conversion of volatile organic acids to methane and carbon dioxide, even though some of these bacteria cannot themselves produce methane. The extended model recognizes as a group (separate from the methanogens) the recently discovered obligate proton-reducing acetogenic bacteria, whose substrates include propionate, butyrate, and higher fatty acids. These compounds are converted to acetate, molecular hydrogen, and carbon dioxide, which in turn are substrates for the true methanogens.

This three-phase model includes in the methanogenic category only those organisms which can themselves produce methane. In digestors they produce methane almost entirely by decarboxylating acetate or by reducing carbon dioxide with molecular hydrogen. Methanogens can produce methane from a limited range of other compounds including formate, methyl-substituted amines, and methanol, but these are quantitatively of limited importance in anaerobic digestion. Thus acetate and hydrogen are important extracellular intermediates in anaerobic digestors, intermediates through which essentially all of the reducing equivalents (COD) present in the original substrate must pass. The non-methanogens in anaerobic digestors must convert this assemblage of organic matter to these simple compounds which the methanogens can degrade; the methanogens cooperate by maintaining these compounds (acetate and hydrogen) at low concentration which facilitates their continued production by nonmethanogens.

Acetate is quantitatively the most important methane precursor, the source of about two thirds of the methane produced in anaerobic digestors. It is degraded by a decarboxylation reaction, in which the carboxyl group is oxidized to carbon dioxide, and the methyl group is reduced to methane, with its substituent hydrogens intact. *Methanosarcinae* and *Methanothrix soehngenii* are responsible for this reaction.

Nearly all the methane produced in anaerobic digestors from sources other than acetate is the result of carbon dioxide reduction with molecular hydrogen; this is the preferred energy-yielding reaction of nearly all the other methanogenic species, and concentrations are kept very low in anaerobic digestors by the continuous removal of hydrogen by these methanogens.

Some acetate-degrading methanogens, including *Methanobacterium soehngenii* and some *Methanosarcinae*, are unable to utilize hydrogen for methanogenesis. For others (e.g., most *Methanosarcinae*), hydrogen is the preferred substrate. However, those methanogens which cannot use acetate show a great affinity for hydrogen, and they may maintain hydrogen at levels so low that acetate-degraders cannot dissimilate it, such that the two methanogenic precursors may each be degraded by separate groups of methanogens. Sometimes hydrogen and acetate are used by the same species: in the studies by McInerney et al., of butyrate enrichments from the rumen, a *Methanosarcina* species was the predominant hydrogen-utilizing species. It was also apparently responsible for acetate degradation, but it was not determined whether the hydrogen and the acetate produced from butyrate were degraded simultaneously or sequentially.

## Interspecies Hydrogen-Transfer

Recent advances in understanding anaerobic digestion can be traced to the resolution of "Methanobacterium omelianskii" into constituent organisms. "Methanobacterium omelianskii" had been reported to oxidize ethanol to acetate, using the generated electrons to reduce $CO_2$ to methane; however Bryant and coworkers found that the culture is not axenic (pure), but rather contains two distinct strains of bacteria, neither of which alone can completely degrade ethanol.

When the organisms are separated, one, a methanogen later named *Methanobacterium bryantii*, grows on hydrogen and $CO_2$ but no other substrates. The other species, S organism, oxidizes ethanol to acetate. The electrons generated from this oxidation are transferred to a pyridine nucleotide carrier (NAD), and ultimately used to reduce protons to molecular hydrogen. Because hydrogen production from ethanol (or from NADH) is thermodynamically possible only at low hydrogen concentrations, S organism requires the removal of hydrogen for complete ethanol conversion. In the case of "Methanobacterium omelianskii" this is accomplished by the action of *Methanobacterium bryantii*.

S organism can grow alone on other substrates such as pyruvate, without the removal of hydrogen by a methanogen. However hydrogen removal is always beneficial, allowing more growth and substrate degradation, and possibly a greater growth yield via the phosphoroclastic reaction. Without hydrogen removal, acetogenic intermediates are diverted as hydrogen sinks to ethanol production.

"Methanobacterium omelianskii" was the first demonstration of hydrogen removal by methanogens affecting the fermentation of nonmethanogens, a phenomenon now known as "interspecies hydrogen-transfer." It is now clear that there are many reactions which can occur in anaerobic digestors, but which are exergonic only when hydrogen concentration is low. Some compounds degraded by these reactions (e.g., ethanol and lactate) are not important intermediates in the anaerobic digestion process, but they may be present in the influent substrate. A number of examples of interspecies hydrogen-transfer reactions were subsequently studied, and fall into three categories: those which degrade small, relatively high-energy compounds; fermentative reactions; and those which degrade very low-energy compounds.

The first, catabolizing small organic molecules such as lactate and alcohols, produces hydrogen and requires its removal by methanogens or other hydrogen-utilizing bacteria. S organism belongs in this group. The bacteria catalyzing these reactions can grow axenically on other substrates (e.g., S organism grows on pyruvate), but catabolize the alcohols or lactate only when hydrogen is removed. In the absence of hydrogen removal, traces of substrate are oxidized and traces of hydrogen accumulate, inhibiting further conversion. Some *Desulfovibrio* species grown in the absence of sulfate fall into this category. With sulfate present as an electron acceptor, compounds such as lactate can be oxidized. When it is absent, protons can replace it, but only when molecular hydrogen is kept at low concentration by interspecies hydrogen-transfer.

The second category of interspecies hydrogen-transfer occurs with fermentative bacteria catabolizing carbohydrates. These bacteria grow well in the absence of hydrogen removal, producing little or no hydrogen. However when hydrogen-utilizing species such as methanogens are present, the fermentative bacteria greatly increase their hydrogen production, often increasing their growth yields as well. When

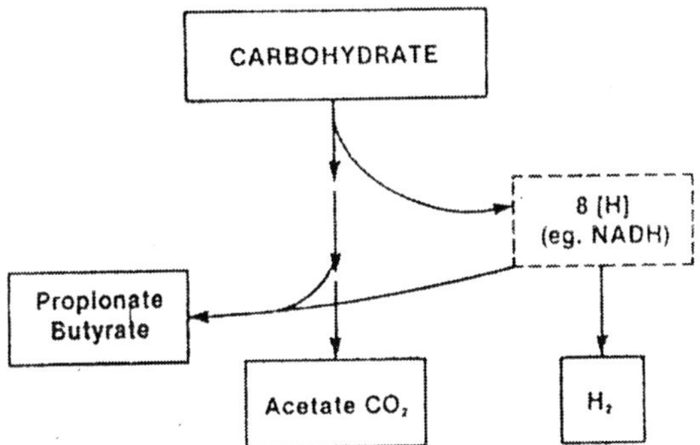


*Fig. 8.2. Fate of electrons on reduced pyridine nucleotides in many fermentative bacteria.*

fermentative bacteria catabolize carbohydrate, electrons are generated and stored temporarily as reduced pyridine nucleotides. These nucleotides must be recycled, by conversion back to the oxidized form, in order for the fermentation to continue. One possible target of the electrons carried on reduced pyridine nucleotides is use by hydrogenase enzymes for proton reduction, producing hydrogen gas. In this case the products are two moles of acetate, two moles of $CO_2$ and four moles of molecular hydrogen per mole of hexose. The thermodynamics of hydrogen production from reduced pyridine nucleotides allow it to proceed only at low hydrogen concentrations. If hydrogen accumulates, this reaction is blocked, and the fermentative bacteria must find other targets for the electrons in order to reoxidize the pyridine nucleotides. This is accomplished by transferring the electrons to glycolytic intermediates, forming reduced products such as lactate, ethanol, propionate or butyrate, at the expense of acetate production. This represents a loss of potential growth-energy for the fermentative bacteria, because with the phosphoroclastic reaction, acetate production is accompanied by the production of ATP.

In the third category of interspecies hydrogen-transfer reactions, hydrogen production is an obligate process. Unlike S organism or *Desulfovibrio* species, bacteria in this group have no alternate substrates with which they can grow in axenic culture, and so are obligate syntrophs, requiring partners which can remove molecular hydrogen. They have been named the "obligate proton-reducing acetogens," and accomplish the conversion of propionic and longer monocarboxylic acids to the methanogenic substrates acetate and hydrogen. The other unique characteristic of this category of reactions is that the concentration of hydrogen required to make them exergonic is much lower than that required for reactions of the other two categories.

The first discovery of bacteria belonging to this group was reported by McInerney, et al. *Syntrophobacter wolfei* oxidizes butyric and longer (up to octanoic) normal monocarboxylic acids, producing acetate and molecular hydrogen (and propionate in the case of odd-chain length acids). It can also oxidize isoheptanoic acid, from which it produces acetate, hydrogen, and isobutyrate. These findings are consistent with a B-oxidation mechanism, although this has not been established. Thermodynamic considerations require hydrogen concentrations below $10^{-4}$ atmospheres for these reactions to occur, and so the organism can function only when co-cultured with hydrogen-utilizing bacteria. It was isolated using lawns of hydrogen-utilizing bacteria (desulfovibrio or methanogens). Colonies in roll tubes take many weeks to develop, and, when picked, contain *Syntrophomonas wolfei* and only one other species, the lawn organism. Monoxenic co-cultures of *Syntrophomonas wolfei* have also been prepared using *Methanospirillum hungatei* or other methanogens as the hydrogen-utilizing species. These co-cultures accomplish the degradation of butyrate in a manner analogous to ethanol oxidation by "Methanobacterium omelianskii." The oxidations of the fatty acids do not occur when hydrogen is present at significant levels or when hydrogen removal by the hydrogen-oxidizing bacterium is inhibited.

*Syntrophobacter wolinii* oxidizes propionate to acetate and carbon dioxide, and, like *Syntrophomonas wolfei*, produces molecular hydrogen only at low concentrations. It was also isolated in monoxenic coculture with *Desulfovibrio* strain G11 as the hydrogen-utilizing species, and grows even more slowly than *Syntrophomonas wolfei* (minimum generation time 3.6 days). The generation times of the bacteria which oxidize propionic and longer volatile organic acids are similar to the minimum solids retention times necessary to maintain populations of propionate- and butyrate-degrading bacteria in anaerobic digestors.

A benzoate-degrading bacterium was also isolated which requires hydrogen removal for growth. It can degrade no other of the tested aromatic compounds, and grows even more slowly than the other obligate proton-reducing acetogens, with a minimum doubling time of 5.5 days. Obligate proton-reducing acetogens are not the only species which degrade aromatic compounds, as some bacteria accomplish this fermentatively.

One interesting class of obligate proton-reducing bacteria has been found to be not acetogenic, but rather acetate degrading. Zinder reported the isolation in co-culture of a bacterium which oxidizes acetate to

$CO_2$ producing molecular hydrogen; this also requires hydrogen removal, and occurs only at low hydrogen concentrations. The organism is particularly interesting in that it degrades acetate by a reaction which conforms to the van Niel's $CO_2$ reduction theory of methanogenesis. Acetate degradation is the main exception to the theory, which predicted that all organic matter is oxidized to $CO_2$ and generated electrons are used to reduce part of that $CO_2$ to methane, so that all biogenic methane is the result of $CO_2$ reduction. Most acetate in anaerobic digestors is degraded by decarboxylation, but these acetate-oxidizers may be of limited importance there.

All but one of the described obligate proton-reducing species are mesophilic. One thermophilic bacterium, in monoxenic co-culture with *Methanobacterium thermoautotrophicum*, oxidizes butyrate to acetate, transferring molecular hydrogen to the thermophilic methanogen.

**Hydrolysis and Initial Reactions of Anaerobic Digestion**

The wide range of compounds which enter an anaerobic digestor must be converted to substrates utilizable by methanogens if it is to be converted to the terminal products methane and $CO_2$. These substrate materials range from insoluble macromolecules, whose degradation may require many different reactions catalyzed by interacting groups of bacteria, to simple compounds such as acetate, which can be degraded directly to terminal products. The nature of the substrates of an anaerobic digestor of course depend on their preparation, transportation, and storage history. Initial degradative reactions may occur before the substrates enter the anaerobic digestor. For instance, very little readily fermentable organic matter (e.g., soluble carbohydrate) is found in domestic sewage sludge, because it undergoes fermentation in the collection system before reaching treatment plant. Animal waste stored for as little as three days can have significant concentrations of lactic acid, although lactic acid is not present in fresh wastes, nor is it produced in healthy anaerobic digestors.

The compounds which actually enter an anaerobic digestor may enter into the generalized pathway for degradation at a number of different places. Polymers, such as cellulose, starch, or proteins, are first hydrolyzed, mainly to oligomers. These oligomers, as well as monomeric carbohydrates, amino acids, and some organic acids (such as lactic) are degraded in digestors by fermentative reactions. Fats are hydrolyzed to glycerol and long-chain fatty acids; glycerol is subsequently degraded by fermentative reactions, but the fatty acids are apparently metabolized by obligate proton-reducing acetogens, entering the metabolic scheme where propionate and butyrate are degraded. Common to the pathway for degradation of these or any other organic compound degraded completely in anaerobic digestors is the conversion to substrates of methanogens.

**Hydrogen Concentration and Effects on Metabolic Pathways**

Molecular hydrogen is an important extracellular intermediate in anaerobic digestors, with about 30 percent of the methane produced by the reduction of $CO_2$ using molecular hydrogen. In addition, hydrogen concentration influences other reactions at several control points. By its action at the three control points, high hydrogen concentration causes increased propionate and butyrate production at the same time their degradation rate is decreased, resulting in their accumulation. Increased hydrogen concentration can occur in digestors due either to inhibition of methanogenesis or to an increase in hydrogen production.

At low hydrogen concentrations, protons are the preferred electron acceptor for excess electrons, resulting in the production of molecular hydrogen as the electron sink, plus acetate and $CO_2$. At higher hydrogen

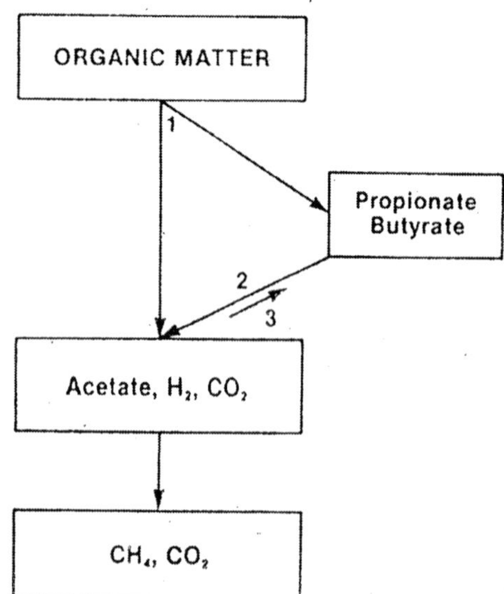


*Fig. 8.3. Hydrogen control-point in anaerobic digestion. The concentration of hydrogen controls the flow of carbon in anaerobic digestors in at least three places: (1) the reaction of organic matter converted to higher fatty acids; (2) the rate at which those higher fatty acids were degraded and (3) the rate of back reaction in which higher fatty acids are produced from acetate and hydrogen (and sometime $CO_2$).*

concentrations, the production of hydrogen from reduced pyridine nucleotides is endergonic, and alternate electron acceptors are required. In digestors this results in increased production of propionate and butyrate. The influence of hydrogen concentration on the production rates of propionate and butyrate has been termed hydrogen control point number 1 by McInerney and Bryant.

The second control point of hydrogen concentration is the inhibition of propionate and butyrate degradation, which is complete and rapid except when hydrogen concentration is very low.

The third location at which hydrogen concentration can influence fermentation patterns in anaerobic digestors affects the production of volatile organic acids (acetate and longer) from hydrogen, acetate to $CO_2$. There are a number of bacteria which have been isolated from digestors and other sources which can grow by using $H_2$ to reduce $CO_2$ or acetate, producing higher volatile organic acids. These reactions do occur in digestors, and bacteria which can catalyze them to obtain energy can be found there, suggesting that conditions for which the reactions are exergonic occur there. These reactions can be thought of as "back reactions," since they are the reverse of those which lead to the terminal products methane and $CO_2$; thus under identical conditions (i.e., concentrations of reactants and products) they will have a free-energy change equal but of opposite sign to the respective forward reactions. Both kinds of reactions can be simultaneously exergonic only if there are distinct microenvironments in the digestor having different concentrations of compounds involved in those reactions. Of the possible compounds, hydrogen has a turnover rate much more rapid than the others, and so is the most likely to have significant concentration gradients in digestors. Hydrogen production from a particulate organic matter may be a point source of hydrogen which may generate these gradients.

Apart from the three control points described above, high levels of hydrogen can influence digestors by inhibition of acetate degradation by acetoclastic methanogens; this effect has been demonstrated in enrichments and pure cultures and in digestors. When substrates from which hydrogen is produced and degraded in mixed-culture systems also containing *Methanosarcinae*, acetate degradation usually starts only after that hydrogen-generating substrate is depleted. (However, if hydrogen remains at very low concentrations during its production, such as when it is produced from butyrate, acetate degradation can accompany hydrogen production.) Formate also inhibits acetate degradation in enrichments, probably because hydrogen is produced from it. If hydrogen is bubbled into digestors, inhibition of acetate degradation occurs. Hydrogen-grown *Methanosarcina barkeri* strain MS has a long lag phase after hydrogen is removed before it can rapidly degrade acetate. Although some strains of methanogens in culture can convert acetate to methane only in the presence of hydrogen, the conversion rate is very small, and is probably not quantitatively important in digestors. Hydrogen does not inhibit cultures which are almost exclusively *Methanothrix soehngenii*; unlike most *Methanosarcinae*, they cannot catabolize hydrogen.

## Factors Affecting Rate and Extent of Methanogenesis

### Rate-Limiting Reactions

For anaerobic digestion of most waste materials, it is hydrolysis of polymers which limits the extent of methanogenesis from organic wastes. With a properly designed and operated digestor, the theoretical potential of methane production represented by the soluble organic matter in the effluent is usually very small compared with the total amount of methane produced. When viewed in this way, the potential for improvement of conventional digestion lies in increasing the extent of hydrolysis of polymers. However the fermentative and methanogenic reactions can limit methanogenesis in two ways: (1) instabilities in the fermentation can cause digestors to go sour, and (2) the slow growth of bacteria accomplishing the conversion of volatile organic acids to methane and $CO_2$ require long solids retention times in order to maintain adequate populations in digestors.

When digestors are operated at high loading rates there is a rapid turnover of volatile organic acids, so that they accumulate rapidly if their degradation is inhibited. When their levels begin to increase, halting substrate addition sometimes allows the acetogenic and methanogenic microflora to recover. However, the continued fermentation of slowly-degrading substrates may cause the increase to continue. The ultimate result of volatile acid accumulation is a drop in pH, which further inhibits methanogenesis. This situation is known as "sour digestion," and it can be expensive and inconvenient.

When very dilute or soluble compounds form the substrate, they are rapidly fermented, and it is often the growth rate of the slow-growing methanogens and acetogens which limits methanogenesis. Various methods have been used to increase the hydraulic loading rate of digestors while maintaining solids (i.e., bacteria) retention time at a level high enough to maintain adequate populations of slow-growing bacteria. The traditional technique is to recycle solids, but more novel techniques are now receiving increased attention. The ability of some bacteria to flocculate or adhere to inert surfaces can be exploited to keep them within the fermentor. The anaerobic filter and related techniques rely on this ability, allowing liquid to flow through the system rapidly while organisms remain within the reactor. This can give hydraulic retention times as short as a few hours while maintaining solids retention times long enough for methanogens and acetogens to florish. The principle of the downflow system is different, but its operating characteristics are similar. Recently immobilization with alginate of the methanogenic and acetogenic bacteria was shown to be an effective method for maintenance of populations in systems with short hydraulic retention times.

**Temperature**

There is some controversy as to whether there exist two temperature optima (i.e., mesophilic and thermophilic) for anaerobic digestion. Pfeffer examined anaerobic digestion of domestic refuse at temperatures between 35° and 60°C at 5°C intervals, and found 40°C consistently more favourable than 35° or 45° when the retention time was between 4 and 30 days. With the same retention times he found a second optimum at 60°C even more favourable than at 40°C. Digestors perform more satisfactorily at 60°C than either 55° or 65°C. There are only small differences in gas production of completely-mixed anaerobic digestors at temperatures between 30° and 60°C when the loading rate is not high (less than 7 g/L) and the retention time is at least six days. However when loading rates are high or retention times short, thermophilic digestors usually outperform digestors operated at lower temperatures.

Buhr and Andrews designed a dynamic process model to describe effects of temperature. This model predicts that the temperature which gives minimum volatile organic acid concentration increases with decreasing retention time so that at very low retention times (3.5 days) the optimal temperature is high (about 50°C). It also predicts greater maximum methanogenic rates at increasing temperatures up to 60°C. These predictions are generally consistent with later studies which tested shorter retention times than were previously investigated (a thermophilic [55°C] cattle waste digest or operated at a 3-day retention time produced more methane than digestors operated at longer [6, 9, or 12 day] retention times, although volatile organic acid levels were higher). Steady-state conditions can be achieved with a retention time as short as 2.5 days, the shortest reported for a completely mixed digestor. Stable conditions at this short retention time were achieved only at 60°C when tested at temperatures in 5°C increments between 30°C and 65°C.

In these studies, the digestors were always allowed a period of acclimation to the temperature at which tests were done. To obtain thermophilic digestion, digestors can be started up at the desired temperature with a low initial loading rate, or temperatures of mesophilic digestors can be shifted slowly (about 1°C per week). When the temperature is raised gradually there is a reproducible decrease in gas production at 43°C, but maximum gas production is again achieved at 45°C. More rapid changes in temperature may not cause ill effects if the digestor is not already operated near maximum loading rate or minimum retention time; animal waste digestors can normally be shifted abruptly from 55° to 60°C, giving increased methane production. Of the culturable bacterial population in a mesophilic digestor, 9 percent can grow at 50° and 1 percent can grow at 60°; temperature acclimation of digestors may then be accomplished by a selection of thermophilic subpopulations at high temperatures.

Little is known about effects of sudden decreases in temperature of digestion (as would result from failure of heating systems) but the model of Buhr and Andrews predicts that a drop from 50°C to 40°C would cause digestor failure within two to three days.

Advantages of thermophilic digestion aside from its ability to operate at shorter retention times and higher loading rates include better destruction of bacterial and viral pathogens and better dewaterability of the sludge. Its chief disadvantage is the heat requirement to attain and maintain the higher digestor temperature.

Temperatures higher than 60°C have consistently been found to give less favourable fermentations. Volatile acids concentrations were higher at 65° than 60°C, but stable fermentations can sometimes be achieved. Pohland and Bloodgood were not able to achieve a stable methanogenic fermentation at 60°C. There have been no detailed investigations of digestion temperatures higher than 65° although one report found no acclimation of the digestor to 65° even after 50 days. It is possible that because the complex nature of anaerobic digestion requires many interacting groups, it is unable to function normally at very high temperatures. In extreme environments such as high temperatures, microbial diversity is known to be limited.

## pH

In general the optimal pH for mesophilic digestors is between 6.7 and 7.4, and they do not function well if the pH is below 6 or above 8. The optimal pH range for thermophilic digestors has not been investigated, but the normal operating pHs for thermophilic digestors tend to be higher. The higher vapor pressure of water dilutes gaseous $CO_2$ giving a lower partial pressure, and $CO_2$ itself is less soluble, giving a lower aqueous concentration of that acidic gas.

The pH in anaerobic digestors is controlled mainly by the bicarbonate buffer system. Other compounds either do not change their extent of ionization greatly over the normal pH range (e.g., volatile organic acids and ammonia are essentially completely ionized at near neutral pHs), or they occur at concentrations too low to function as effective buffers (the pKa's of NaH2PO4 and H2Saq are close to operating pHs of anaerobic digestors, but they do not function significantly as buffers because the bicarbonate concentration is so much higher). Therefore the pH of digestors is controlled by the partial pressure of $CO_2$ in the produced gas and the concentrations of acidic and basic substances in soluble intermediates or the added substrate.

Acidity in anaerobic digestors comes predominantly from aqueous $CO_2$ dissociation and from volatile organic acids, which dissociate. One souce of basic compounds in digestors is ammonia, produced from proteinaceous material in the digestor or present in the substrate added to it. This can be an important source, especially when the C:N ratio is low. Other basic substances can occur in the substrate, in the form of inorganic salts or salts of organic acids. The total of these basic substances must balance the concentration of bicarbonate (from $CO_2$ dissociation) and volatile organic acid salts in the digestor to achieve pH stability.

If volatile organic acids are produced by the fermentative bacteria more rapidly than they are removed by the obligate proton-reducing acetogens and methanogens, the acids will accumulate. If the problem is not resolved, a drastic drop in pH will occur, inhibiting the methanogens and exacerbating the situation. This can be a significant problem in commercial digestors, and generally addition of substrate is halted until recovery. There has been some debate about whether the low pHs or the acids *per se* are inhibitory to methanogens, and whether adding alkaline materials to raise the pH is effective in enhancing recovery. McCarty and co-workers showed that the acids are not inhibitory, but that the cation of the alkaline compound added could be inhibitory (e.g., NaCl is as inhibitory as sodium acetate). They also showed that a proper balance of monovalent (Na and K) and divalent cations gives minimal inhibition of methanogenesis.

**Stoichiometry**

When molecules containing only the atoms, C, H, and O are fermented to the gasses $CH_4$ and $CO_2$, stoichiometric calculations of the theoretical products can be made using the following formula

$$C_nH_aO_b + \left(n - \frac{a}{4} - \frac{b}{2}\right) H_2O \rightarrow \left(\frac{n}{2} - \frac{a}{8} + \frac{b}{4}\right) CO_2 + \left(\frac{n}{2} + \frac{a}{8} - \frac{b}{4}\right) CH_4 \quad ...(1)$$

The equation can be modified for other atoms in fermentations if the reaction path is known. For instance, sulfur present in the organic matter is converted to hydrogen sulfide, and to alter the equation to include this, the equation would become:

$$C_nH_aO_bS_c + \left(n - \frac{a}{4} - \frac{b}{2} + \frac{c}{2}\right) H_2O \rightarrow \left(\frac{n}{2} - \frac{a}{8} + \frac{b}{4} + \frac{c}{4}\right) CO_2 + \left(\frac{n}{2} + \frac{a}{8} - \frac{b}{4} - \frac{c}{4}\right) CH_4 + cH_2S \quad ...(2)$$

These stoichiometric equations, however, do not take into account the dissociation of carbon dioxide to bicarbonate. The "$CO_2$" in the equation refers to all forms of carbon dioxide, including gaseous, aqueous, bicarbonate, carbonate, and insoluble salts. The amount of this "$CO_2$" which leaves the digestor with the effluent will depend on two factors, the rate at which effluent leaves the digestor and the concentrations of dissolved and precipitated species. So digestors having a short hydraulic retention time would tend to have more $CO_2$ leaving in the liquid phase, and a higher proportion of methane in the gas. The concentrations of dissolved species depend on the partial pressure of $CO_2$, its solubility, and its extent of dissociation. The solubility and extent of dissociation are determined by pH, temperature, and ionic strength.

The equations above can be used with complex wastes if the elemental composition is known; the empirical formula is then used in the calculation. Also in this way the effects of inorganic electron acceptors present in the substrate can be seen on the products. When significant quantities of inorganic electron acceptors such as sulfate, nitrate, or oxygen enter the digestor, they can change the stoichiometry of the reaction. If their quantities and their products in the fermentation are known, they can be figured into the stoichiometric equation by modifying the first term of the equation above to include them in an overall empirical formula for the substrates of the digestor. For instance if glucose is the substrate, but oxygen concentration is 10 percent of the glucose concentration, the empirical formula of the substrate would then be $C_6H_{12}O_6$ + $0.1O_2$ or $C_6H_{12}O_{6.2}$. These equations assume that the substrate is completely converted to $CH_4$ and $CO_2$ and so should be applied only to the biodegradable portion of the organic matter.

Also a portion of the biodegradable organic matter will be converted to microbial cells rather than converted to methane and $CO_2$. Although this portion varies depending on the energy content of the degraded organic matter, it is roughly 10 percent of the total.

## Nutrition and Inhibition

Nutritional requirements for anaerobic fermentative bacteria usually include a fermentable carbohydrate or amino acid, minerals, and often some B vitamins. As in many other aspects of anaerobic digestion, much information about bacterial nutrition is inferred from isolates of the rumen ecosystem, which has been more thoroughly studied. The differences between those ecosystems should be considered when making these inferences, however. The organic loading rate of the rumen is much higher than most anaerobic digestors, the retention time of liquid and solids is shorter, and the concentrations of volatile organic acids and other compounds which may function as growth factors are higher. An environment, such as the rumen, where these potential growth factors are present in high concentrations tends to have more bacteria which require them than does an environment which does not have as high levels of these growth factors (such as digestors). The "minimum-requirement mechanism" described by van Niel indicates that, for instance, bacteria having a nutritional requirement for a compound available in the rumen would tend to predominate there over similar organisms which did not have that requirement. If digestors did not have the required concentration of that growth factor, the rumen species would be outcompeted there by bacteria which did not have that growth factor requirement, and rumen species may tend to have more nutritional requirements than digestor species.

Phosphate and iron are normally essential for anaerobic digestion, while nickel, cobalt, and molybdenum are required for some methanogenic bacteria. Nickel is required for factor F430, a coenzyme of methanogenic bacteria, although very high levels of nickel have been shown to be more inhibitory to methanogenic bacteria than to fermentative bacteria. The very high levels of iron addition required to give maximal levels of methanogenesis in biodegradability assays was probably not due to requirements for high concentrations of iron, but rather due to traces of nickel in the iron solution which satisfied the nickel requirement only when large amounts of iron were added. Iron could also be stimulatory by precipitating phosphate, which otherwise could precipitate other required trace minerals.

Other metals, especially calcium, magnesium, sodium, and potassium, are often required as well. The concentrations of metals relative to each other and to certain other metals, especially heavy metals, are as important as the absolute concentrations; in this respect they can be considered as inhibitors as well as nutrients. The levels of phosphate and sulfide can also effect the concentrations of these metals by precipitation. Sulfide is the normal sulfur source for bacteria in digestors, although some species require or are stimulated by cysteine or methionine, and a few can use sulfate. Ammonia is required as a nitrogen source by many species, but amino acids or short polypeptides are stimulatory to some, and required by a few. In addition, some species can use urea or nitrate as a nitrogen source. Some rumen species are stimulated by high levels of acetate, although for many the carbon sources for growth can be the intermediates of the fermentation. Some require other volatile organic acids, especially isobutyric, isovaleric, and 2-methylbutyric acids. These, as well as butyrate and valerate, may be required for fatty acid or amino acid synthesis. In some rumen strains $CO_2$ is required as a carbon source, and in some it is also necessary in dissimilatory reactions which produce succinate from pyruvate.

When nutrients are present at concentrations much higher than required they can become inhibitory. High concentrations of ammonia are inhibitory, although digestors can become acclimated to them. Sulfide can also be inhibitory. The importance of mineral nutrients in anaerobic digestors can be studied with batch or continuous culture experiments to determine concentrations which give optimal rates of methanogenesis.

For other nutrients, individual bacteria grown in pure culture may have requirements which are normally met in the digestor by cross-feeding from other species, and so are not required as additions to the digestor. These nutrients, such as several B- and other vitamins should be included when axenically culturing bacteria from digestors, although their addition to digestors may not be stimulatory.

When animal wastes are digested there may be antibiotics present at inhibitory concentrations. These can cause digestors to fail, although acclimation of the microflora to these antibiotics can occur. Both mesophilic and thermophilic digestors have been demonstrated to adapt to monensin or chlortetracycline containing animal wastes if they are given time to acclimate.

# 9

# CONTROL OF FOREST PESTS

Since 1962, with the publication of *Silent Spring* by Rachel Carson highlighting the adverse effects of pesticides on birds, there has been renewed interest in biological pest control. Biological control is based on two ecological principles: that one organisms can be used to control another organisms; and that some of the organisms that can be used to control other organisms have a limited host range. Since this approach is close to nature's own way of controlling an organism, it has found widespread acceptance with the public. Biological control is perceived, for good reasons, as being a highly selective and effective method that is perceived, for good reasons, as being a highly selective and effective method that is long lasting. At the same time, some organisms have been used for practical or commercial purposes such as using yeast to make bread or beer since ancient times, and this use has been defined as 'biotechnology'. Beginning in the 1970s, biotechnology has taken a dramatically different turn. With the knowledge gained in recombinant DNA technology, it became possible to alter the genetic make-up of an organism by inserting foreign genes, deleting genes, or modifying the functions of genes. The integration of traditional biological control with modern biotechnology has resulted in some exciting new developments, some of which will be outlined in this chapter.

Modern biotechnology is an immensely powerful tool and has found applications in almost all walks of life, promising improved agricultural and forest products, health care, and crime detection to mention a few. Unfortunately, research in biotechnology has become controversial, mainly because it is difficult for the general public to comprehend this field since it is mired in complex jargon. It has been cited as being a threat to environmental safety, violating natural laws, and religious zealots have even labelled it as trivializing the meaning of life! In this chapter we will examine some of the recent advances in pest control using biotechnology and attempt to allay some of the concerns of the public.

## Pests of Forests

Broadly speaking there are three major groups of forest pests: pathogenic micro-organisms, competing vegetation, and insect pests. Historically many of these pests of economic importance have been managed by using chemical control agents such as chemical fungicides, herbicides and pesticides. While biological control methods have been attempted, there was no sustained interest until recently when some of the non-target effects of the chemical pesticides became apparent. After a cursory examination of the biological control of competing weeds and pathogenic organisms, much of the discussion will be focused on the control of insect pests.

### Vegetation Management

Phytophagous biological control of organisms have been tested as weed control agents for many decades; the organisms used have been either insects or microorganisms, including fungi, nematodes or mites. Testing for host specificity has been a major hurdle. In Australia, the common heliotrope Heliotropium europaeum was successfully controlled with the fungus Uromyces heliotropii and tests were conducted on 96 plants of the region for microscopic and macroscopic effects. Rather than continuing to depend on chemical herbicides, North American forestry is rapidly changing to 'ecosystem management', which is an effort to combine ecological principles, sustainable forests, and land stewardship ethics. Since chemical herbicides are generally inexpensive and effective there has been a reluctance to invest in research on the use of alternative herbicides. Public belief that forested landscapes are among the only remaining unspoiled eco-systems has more recently lead to a search for other types of weed control agents including mycoherbicides. Chondrostereum purpureum is a common pathogen that invades the xylem vessels through fresh wounds and causes silverleaf disease in various hardwood shrubs and trees. Although *C. purpureumis* not host specific, it prefers broadleaf trees. The fungus lives in living tissues or in trees that have been dead for less than 2 years. The airborne basidiospores infect tree wounds or strumps.

Biotechnological approaches have been used to engineer herbicide resistance genes into the trees to protect them against herbicides that are sprayed to control competing vegetation. A sulphonyl urea herbicide, chlorosulfuron, was shown to inhibit cell division. Biochemical studies showed that it inhibits acetolactate synthase (ALS), which is required for the synthesis of isoleucine, leucine and valine. Using the ALS gene as a probe, the ALS mutant gene from herbicide resistant plants was isolated. The resistance was conferred by selective amino acid substitution. When such a resistant ALS gene was cloned into the tobacco plant, the transgenic tobacco plant was found to be resistant to sulphonyl urea herbicides. A similar approach was used to isolate a glyphosate resistant gene that was transferred into the crop plant. In the case of phosphoinothricin, a gene that produces a detoxifying enzyme was found to be better than a resistant glutamine synthase gene. In the case of atrazine and bromoxynil, the herbicide binding protein or $Q_\beta$ protein (32-kDA membrane protein in the chloroplast) was involved in the resistance mechanisms. The resistant variety of the $Q_\beta$ protein has some amino acid substitution.

**Plant Pathogens**

Many species of bacteria, fungi and nematodes are pathogenic to plant in general. The molecular mode of action of phytopathogenic bacteria is being actively studied and the results of these investigations may pave the way for developing resistant varieties by recombinant DNA technology. *Erwinia*, *Pseudomonas* and *Xanthomonas* have genes that code for cellulases (endo-β-1,4-glucanase) but their role in pathogenicity has not been established. Proteases, productio n and export of extracellular enzymes, polysaccharides, plant growth substances, toxins and unknown products have all been implicated in pathogenicity and genes for many of them have been clones. *Agrobacterium tumefaciens*, the soil bacterium that causes crown gall in dicotyledonous plants, has been far the most studied and is extensively used for gene transfer. During the disease process, a segment of this bacteriums's DNA (T-DNA) is transferred to the host plant and becomes integrated into the plant genome. The T-DNA originates from a 200-kb plasmid and foreign genes can be inserted into this DNA for transfer into the host plant.

The molecular biology of several fungal pathogens of plants has been well studied. A resistance gene may be cloned by a short-gun method and thereby inserted into a host plant. The large genome size and, consequently, the size of the cloned DNA fragments however requires the screening of more than 10 000 transformed plants, which is an enormous task. Another method is the insertion of a transposable DNA sequence (transposable element) which in essence functions as a tag and can carry a resistance gene.

**Nematodes**

Resistance to phytopathogenic nematodes has been reported and has been traced to an H1 resistance gene in certain cultivars of potato. While at this time this resistance gene has been shown for the potato cyst nematode, Globodera rostochiensis, similar studies can be extended to others that are pathogenic to tress such as the pinewood nematode. The excellent genetic information available for the free-living nematode Caenorhabditis elegans will serve as an elegant backdrop to characterizing the biochemical mechanism of resistance in parasitic forms such as the potato cyst nematode. Since the H1 resistance gene has been shown to follow a Mendelian pattern, it lends itself to genetic manipulation.

**Phytoalexins**

Phytoalexins are low molecular weight, antimicrobial agents that are synthesized by plants in response to a pathogen or stress factor. Many compounds from the invading fungus or bacterium, such as oligoglucans, ethylene, chitosan oligomers, andpolypeptides serve as elicitors for phytoalexin production in the host. The accumulation of phytoalexins at the infection site inhibits the growth of the fungus or bacterium and serves as a defence mechanism. The ability to produce phytoalexins confers a degree of resistance against the pathogen. Work that will lead to gene expression for phytoalexin accumulation and disease control is underway.

**Viral Pathogens of Plants**

Viral pathogens of plants are well known and transmission of many of them by sucking insects such as aphids and leafhoppers has been extensively studied. One of the novel ways of providing protection against the pathogenic viruses is similar to vaccination. The coat protein of the virus protects the host plant from viral infection. The exact mechanism for this protection has not yet been elucidated. For example, the cucumber mosaic virus (CMV) infects 775 species of plants from 85 different families and causes severe viral disease. The coat protein (CP) gene of this virus was inserted into tobacco leaf discs using Agrobacterium tumefaciens. Of the 22 transformant lines, 15 showed resistanc to CMV. Such a technology may be extended to trees for protection against viral pathogens.

## Insect Pests

Chemical and biological control of insect pests using parasites and predators dates back to the 1950s and has been extensively reviewed. Control of forest insect pests with the introduction of various egg parasites (*Trichogramma* sp.), external parasites (Tachinids), and internal parasites (Ichneumonids, Chalcids, etc.), has been attempted with various degrees of success, but in general, the result have not been spectacular. However, success stories of classical examples such as the introduction of the Australian vedalia lady beetle in 1899 to control the cottony-cushion scale in Californian orange groves reinforced the need to continue this approach. One of the areas that needs particular attention, especially in forestry, is the environmental impact of classical biological control agents. In the words of Howarth, 'Absence of evidence is not evidence of absence'.

Microbial pathogens such as fungi, bacteria and viruses that cause diseases in insects have been tested extensively as biological control agents. Among the fungi, more than 50 genera contain species that are pathogenic to insects and are candidates for biological control. They have widely varied life cycles and host range and are transmitted primarily through either spores (Sexual forms) or conidia (asexual forms). When a spore or a conidium lands on a susceptible host, it adheres to the cuticle, germinates and digests it entry into the insect by means of its germ tube. The fungal mycelia soon overpower and mummify the insect. The conidia or spores are released and the life history starts all over again. *Beauveria bassiana* is an imperfect fungus that infects over 100 insect species from many orders; it has been used extensively in Russia and China for insect control. Identification of the various isolates (pathotypes) that are morphologically similar is being done with molecular techniques to resolve taxonomic problems. In addition, entomophthoraceous fungi have been reported from many insects, and many of them have been found to have a narrow host range. Recently, recombinant DNA technology has been used to improve the efficacy of some of the mycoinsecticides. Additional copies of a gene encoding a regulated cuticle-degrading protease from an entomopathogenic fungus, Metarhizium anisopliae, were inserted into the same fungus so that the enzyme was constitutively overproduced in Manduca sexta infected with the recombinant fungus. The insects are less, died quickly and the cadavers were black due to melanization. The resulting cadavers were poor substrates for sporulation ensuring minimal persistence of the genetically altered fungus.

Phylum Protozoa is now considered a sub-Kindgon, and the class Microspora has now been elevated to a phylum and includes the entomopathogenic microsporidia; this is important from an insect control stand point. *Nosema locustae* has been registered for controlling grasshoppers. Microsporidia are obligate intracellular pathogens and the larvae are infected when they ingest the spores. The spore extrudes a hollow polar filament through which the sporoplasm is injected into the midgut cells where they multiply vegetatively. Many forest insect pests have micro-sporidia and some of them have been shown to be transovarially transmitted. They are quite infective but lack the virulence required to compete with other control measures. Their large genome has been an impediment to developing recombinant DNA techniques to increase their virulence by inserting foreign genes.

The bacterium *Bacillus thuringiensis* (Bt) is the most widely used pathogen for insect control in North America. Bt is a common organism that is found in the soil, insects, frass, grain dust and leaves. It was first isolated in 1902 as the causative agent for 'Sotto disease' is silkworm in Japan. It is a Gram-positive, aerobic, spore-forming bacterium, which produces a characteristic parasporal crystal that contains the insecticidal toxins. The insecticidal parasporal proteins are activated by the midgut proteases and the activated toxin binds to a specific receptor on midgut cells of susceptible insects. The lysis and destruction of midgut cells eventually leads to the death of the host. The genes that encode the insecticidal proteins are typically located on large transmissible plasmids that determine the host specificity of the isolate or serotype. Bt has more than 50 subspecies and produces a variety of crystal (Cry) proteins. These Cry proteins are classified according to their activity spectrum and molecular mass. Cry I and Cry II are active against lepidoptera and are 135 and 65 kDa, respectively, in size. The Cry III protein of 65 kDa size is active against coleoptera such as the Colorado potato beetle, *Leptinostarsa decemlineata*. Cry IV proteins range in size from 65 to 135 kDa and are effective against mosquitoes and blackflies. The presence of these toxin genes in the plasmids made it relatively easy to isolate and purify them, and in the 1980s most of them were cloned. These toxins have been engineered into insect pathogenic viruses, which is described later, as well as into plants and trees.

**Viral Pathogens of Insects**

Insect viruses causing epizootics have been recognized for namely a century and one of the earliest such occurrences described in North America was in 1913 by Glaser and Chapman on the wilt disease of the gypsy moth, Lymantria dispar, caused by a nuclear polyhedrosis virus (NPV). Perhaps one of the most dramatic controls ever reported was the control of the European pine sawfly, Neodiprion sertifer, in

Ontario, Canada by an NPV isolated from dead larvae sent from Sweden. Various types of viruses are pathogenic to insect but one group, the Baculoviridae, is uniquely pathogenic only to insects. By far the largest group of insects affected by viruses is the Lepidoptera. Baculoviruses and cytoplasmic polyhedrosis viruses form the two groups of viruses that are infectious to insects.

The host range of entomopathogenic viruses varies widely but generally it is quite narrow. Viruses such a the multicapsid nuclear polyhedrosis virus of the beet army worm, *Spodoptera exigua* (SeMNPV) are species specific whereas the virus of the alfalfa loopepr, *Autographa californica* (AcMNPV)) infects and replicates in at least 33 lepidopterous species from seven different families. However, most of the viruses have an intermediate host range and are best described as genus specific, as is illustrated by HzMHPV that infects several species of *Heliothis*, including *H. zea*. Many baculoviruses have been developed commercially or used in a practical manner for controlling specific pests. The effective dosage is generally established in terms of the number of polyhedra or occlusion bodies (OB) required to affect 50 per cent of the treated insects ($LD_{50}$). Since the time taken to have an effect has an important bearing on the ability to protect the crop from damage, the $LT_{50}$ or time taken to kill 50 per cent of the treated insects, usually in days, is also studied.

Larvae typically become infected by NPV when they ingest the polyhedra on the foliage. The polyhedra dissolve in the midgut and the virions are released. These virions fuse with the microvilli of the midgut cell and go through the cytoplasm into the nucleus. The DNA unravels and the virions begin to replicate and give rise to an intranuclear region referred to as the virogenic stroma. Envelope formation and polyhedron synthesis results in the production of occlusion bodies. Some of the non-occluded virions (NOV) come of the cell as budded viruses that infect more cells. Ultimately, when the cells break down and the entire insect liquefies, the occlusion bodies are released to start the cycle once again.

The degree of susceptibility varies from species to species to species. Those insects showing low susceptibility will permit limited infection to only midgut cells, whereas others that are highly susceptible show infection to virtually all tissues. What confers host specificity has largely been unknown until recently when it was shown that specific regions of the DNA are necessary. Maeda et al. (1993) showed that a recombinant NPV containing portions of AcMNPV and portions of

*Bombyx mori* NPV (*Bm*MNPV) can infect *Bombyx mori* cells, but AcMNPV is not infectious to these cells. Host shifts have been accomplished within closely related hosts and similar viruses. When cabbage looper larvae were fed large doses of the tussock moth NPV over 12 generations, it was possible to obtain a virus that would infect the cabbage looper.

Virus control of insect pests is attractive because it is lasting, highly selective and effective. However, to compete against chemical control agents, they have to act fast, be more virulent and cheaper to produce. The advent of recombinant DNA technology has made it possible to make the virus more virulent and fast acting. Since baçuloviruses are propagated in eukaryotic cell lines (i.e., from insects), the expression of any foreign eukaryotic genes inserted into the baculovirus follows the eukaryotic pathway. All the post-translational processing such as phosphorylation, glycosylation, amidation, protein folding, etc., take place ensuring that the product is functional. It is for this reason that *Ac*MNPV has become the system of choice for expressing many eukaryotic genes that produce pharmaceutical products such as inteferon.

## Modified NPVs for Insect Control

Various types of genes from different sources have been inserted into the NPV to improve its virulence. The market potential for genetically modified NPVs as biocontrol agents is immense and only the surface has been scratched so far. It is now theoretically possible to develop transgenci viruses that are not only host specific but also as effective as conventional control agaents.

### *Structure, Biology, and Ecology*

Before embarking on genetic modification of the virus, it is essential that the biology of the virus is well understood. Fortunately, there is a fund of information in this area.. The spruce budworm virus, Choristneura fumiferana, multicapsed nuclear polyhedrosis virus (*Cf*MNPV) is one of the forestry-related viruses that has been well studied. The infectivity, host-range, $LD_{50}$, $LT_{50}$ and ecology have to be studied to provide the baseline information for assaying the activities of the transformed viruses.

### *Biochemical characterization*

NPVs from different insects have been characterized to various degrees. The relatively small size of the circular dsDNA has made it possible to elucidate the structure. The genomes of the *Ac*MNPV and

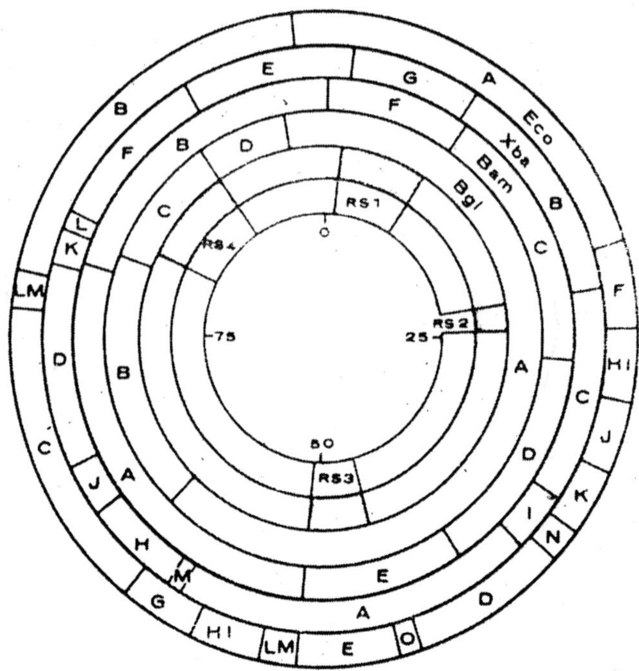


*Fig. 9.1. Circular map with the zero position at the junction of BamH1 fragments B and F.*

BmMNPV have been completely sequenced. The *Cf*MNPV has been mapped and some of the regions have been sequenced. Subsequently, using homologous probes, several genes such as polyhedrin, egt, and p10 have been identified and in many instances their functions have been described. The secret of success for genetic improvement of a virus depends *inter alia* on the availability of a cell line that grows well and permits rapid replication with no loss of fidelity but with retention of infectivity.

### *Development of cell lines*

So far 26 continuous cell lines (1 embryonic, 4 midnight, 11 neonate larval and 10 ovarian) have been developed from the spruce budworm tissues. Most of the cell lines were found to be highly susceptible to the virus and were used as experimental material for work on *Cf*MNPV. A plaque assay was developed using IPRI-CF-124T cells for purification of *Cf*MNPV. Initially, the cell lines were developed in Grace's insect tissue culture medium supplemented with 10-15 cent fetal bovine serum (FBS). Since FBS is expensive, a search for a substitute especially for mass-culturing of cells for virus production resulted in identification of a cell line, CF-203, which grows well in Insect-Xpress medium supplemented with as low as 0.5 per cent serum and permits *Cf*MNPV replication.

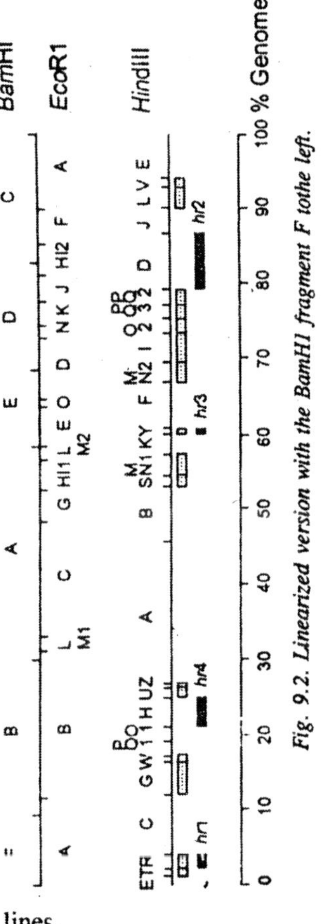


*Fig. 9.2. Linearized version with the BamH1 fragment F tothe left.*

### *Modification of Baculoviruses*

The strategy is to introduce a foreign gene into the viral genome so that the resulting transgenic virus is more virulent and fast-acting than the wild type. Once a candidate foreign gene has been identified, it is inserted into a vector for transferring into the virus. This vector has to be constructed so that it efficiently transfers the foreign gene into the virus. The virus and the vector or recombinant plasmid are added together, or cotransfected, into permissive cells that allow homologous recombination resulting in the production of a few recombinant viruses. These recombinant viruses are then purified from the wild-type virus using a technique called plaque purification. The transgenic viruses can be propagated in the host insect larvae or permissive cell lines.

### *Promoter selection*

One of the primary requirements for producing a successfully modified virus for pest control is the incorporation of the proper promoter that will direct the expression of the foreign gene at the appropriate time and place. For instance, a fast-acting neurotoxin has to be secreted quickly into haemolymph close to the nervous system. Several late gene viral promoters such as polyhedrin and p10 have been used successfully. Large quantities of mRNA for genes expressed under these very late gene promoters will be produced, beginning 24 hours after infection. Polyhedrin promoter is the most commonly used promoter. The advantage of using polyhedrin promoter in a virus where the polyhedrin gene has been removed is that the modified virus is produced without the polyhedrin in a non-occluded form and therefore does not survive in the environment beyond a generation of the host insect. This feature is attractive to both environmentalists because it does not persist in the environment, and to commercial producers because the virus has to be produced every year for repeated application. The disadvantage of using such a system is that there is no reliable quantitative bioassay to assess the effectiveness of non-occluded virions. Currently, the assay is performed by either injecting the modified virus, which can be time consuming, or by feeding the preoccluded virus which is difficult to quantify. Many researchers have been persuaded to use the p10 promoter for the expression of foreign genes because of the difficulties encountered with non-occluded virus particles. The main disadvantage with the p10 promoter is that the modified virus will produce polyhedrin occlusion bodies that can persist and survive in the soil for a long time. Other promoters such as the combination of early and late promoters of baculoviruses and constitutive promoters such as the insect actin promoter that are active all the time have been tried, but none have enjoyed the success of the polyhedrin promoter.

***Candidate foreign genes***

The candidate genes that have been inserted into baculoviruses for improving their control potential; the most promising one are those showing either the deletion of ecdysone glucosyl transferase ($egt^-$) or the expressionof scorpion toxin, mite toxin or juvenile hormone esterase (JHE). The $egt^-$ virus has been one of the leading candidates for early registration because of its acceptance by the public. There is no foreign gene inserted into the $egt^-$ virus and by deleting this gene from the virus the moulting of the host larva is no longer prolonged, which resutls in less feeding damage. Other modified viruses such as the one expressing the scorpion toxin and the enzyme, JHE, are at various stages of evaluation. There are various developmental genes that have a regulatory function and may be tested in the future as candidates for insertion.

***Future consideration***

Producing a genetically altered virus is at the early stages of development at present and there are several hurdles to be crossed before the modified virus can be successfully applied in the field. Some of the major problems that need to be solved are mass production, stability of the modified virus in the field, effect of modifications on the host range, effect of modified virus on non-target organisms, and formulation of the modified virus for field application. Mass producing a fast-acting trangenic virus in an insect host will be difficult because the virus will kill the insects before enough virus is produced. Strategies such as propagating the virus in cell cultures and using an inducible promoter than can be turned on at will by adding the inducers at the appropriate time will have to be employed. Mass production of recombinant viruses in SF-21 cells using large-scale fermenters appears promising. Most of the tests conducted on the modified viruses so far have indicated that the modifications have not affected the host range of the transformed virus and most of the non-target beneficial insects are not affected. The occlusion bodies (OB) of baculoviruses are stable in the environment, especially in the soil, for several years. To circumvent this problem, Wood and his colleagues at the Boyce Thomson Research Institute have been working on a procedure of using a polyhedrin minus (null) preoccluded virus preparation for field application. They have shown that such a polyhedrin null virus disappears rapidly from the soil.

A successful modified baculovirus is one tat the can outcompete a conventional chemical insecticide in its ability to kill faster and be produced at a competitive cost. Some of the promising new strategies include using multiple genes for not only adversely interfering with feeding but also derailing the normal development of the insect past. The receptors and transcription factors that are involved in insect development will become one of the main target areas that will be used for adversely interfering with the normal developmental process. Integrating the rapidly developing field of biotechnology with insect molecular biology, endocrinology, physiology, ecology, cell culture and virology should make the dream of making an ideal recombinant virus a reality in the not too distant future.

**Alternative ways of Improving Baculoviruses**

The traditional approach for improving baculoviruses is to modify the virus by either deleting an existing gene or adding a foreign gene into the viral genome. However, there are alternative ways of improving the baculoviruses as pest control agents. Selecting the more effective strains from the mixture of strains that occur in nature and optimizing the combinations of the more active strains is one method. Another approach is to alter the host range of an effective virus such as *Ac*MNPV so that it can infect other important pests such as the gypsy moth and the sppruce budworm.

***Starin selection and mixed strains***

The baculoviruses isolates from nature contain a mixture of genetically heterogeneous populations. For example CfMNPV field isolates contain at least two different viruses, and one of the two is more infectious to spruce budworm larvae. Two variants were isolated from field-collected *Pf*MNPV. A strain of coding moth granulosis virus (GV) that was 5.6 times more resistant to artificial ultraviolet light in the laboratory and survived twice as long as the wild type in the field was selected for development. Replication of *Ac*MNPV is the presence of 2-aminoppurine resulted in a strain that had increased virulence.

***Host range alteration***

*Ac*MNPV is one of the most virulent and well-characterized baculoviruses. Unfortunately though, *Ac*MNPV does not infect some of the important pests such as the gypsy moth and the spruce budworm. When gypsy moth, *Lymantria dispar,* IPLB-Ld652Y (Ld652Y) cells were infected with *Ac*MNPV alone the infection was enhanced. An *Ld*MNPV gene host range factor 1 (*hrf*-1) that enabled *Ac*MNPV to infect Ld652Y cells was isolated, cloned and inserted into *Ac*MNPV. Modified *Ac*MNPV expressing the *hrf*-1 gene can successfully infect *L. dispar* (Thiem, personal communication). CF-203 cells inoculated with *Ac*MNPV showed a DNA ladder and morphological changes such as plasma membrane granulation, blebbing and nuclear fragmentation, which are characteristic of apoptosis. The mRNA for the apoptosis suppressor gene p35 was detected 9 hours later in *Ac*MNPV-inoculated CF-2003 cells than in SF-21 cells. Only a trace amount mRNA for the *Ac*MNPV-inhibitor of apoptosis homologue (*Ac-iap*) gene and no mRNAs for the late genes, *Ac*MNPV-polyhedrin (*Ac-poth*) and *Ac*MNPV-p10 (*Ac-p10*), were detected in *Ac*MNPV-inoculated CL-203 cells. Inoculation of CF-203 cells with *Cf*MNPV at least 12 hours prior to inoculation with *Ac*MNPV prevented apoptosis-like cell death and mRNAs for *AC-iap*, *Ac-poth* and *Ac-p10* genes were expressed, resulting in successful virus replication and OB production. Work is currently in progress to identify the gene(s) in *Cf*MNPV that provide this protection and enhances *Ac*MNPV replication in CF-203 cells. Once this gene is identified, it should then be possible to engineer a recombinant *Ac*MNPV bearing this *Cf*MNPV gene similar to the recombinant AcMNPV bearing the *Lymantria dispar* MNPV host range gene *hrf-1*. Such a recombinant *Ac*MNPV would most likely infect the spruce budworm and also will probably be more virulent than the wild-type *Cf*MNPV for this forest pest.

## Other Applications of Biotechnology for Pest Control

The recombinant DNA technology used for altering the effectiveness of the virus can also be used on the insect. Incorporation of the genes into the host insect and its parasites and predators can be used in control strategies.

### *Transgenic insects*

Production of transgenic insects carrying an insecticidal gene under the control of an inducible promoter has been under investigation for several years. In this strategy the pest insect is transformed by incorporating a gene that is expressed under the control of an inducible promoter. The transformed insects are mass-produced and field-released to mate with natural populations. After a few generations when the population of the pest insect reaches damaging levels the inducers are sprayed. The insecticidal gene is thereby induced resulting in the death of the pest species. This strategy is similar to the sterile male release technique and can be used in an integrated pest management strategy. Due to the lack of a good procedure for germ line transformation for pest insects this approach has not been very successful so far. With the recent cloning and characterization of transposons from several key pest insects it may be possible to produce transgenic insects in the near future.

### *Improvement of parasites and parasitoids*

There have also been attempts to produce transgenic parasites and predators expressing genes that can enhance the survival of these organisms in the environment. Parasites and predators expressing insecticide resistance genes will work better in the integrated pest management strategies since they will survive insecticide treatments. This approach has had limited success due to the lack of availability of transformation protocols that can be used for producing transgenic parasites and predators. With the recent discovery of transposons in many insects this situation will soon change.

## CONCLUSIONS

Biological control agents in general are the best option for controlling pests in an environmentally friendly manner. Unfortunately however they do not always perform as efficiently as chemical control agents. Furthermore, it costs more to produce biological control agents than conventional chemical pesticides. The advent of recombinant DNA technology has changed the entire situation. It is now possible to enhance the control potential of biologicals by genetic manipulation. We can not only increase the activity to match chemical pesticides but also in many instances retain the host specificity. We are on the threshold of developing new, genetically modified organisms that will revolutionize the concept of pest control. We can now make 'designer pesticides' that will be able to control many pests with surgical precision with little or no impact on non-target organisms. New concepts and new techniques often come with their own set of problems. Establishing their safety educating the users, and perception by the public are some of the aspects that will ultimately decide the fate of this new generation of genetically engineered transgenic biocontrol agents.

# 10

# Drugs from Plants

Plants have been used by man since the beginning of human culture for a great variety of purposes, including medicine. The earliest known record of a plant being used in medication is found on an Egyptian papyrus dated about 1550 BC. Since then, plants have provided nearly half of the world's successful drugs, ranging from anticancer drugs from the tiny periwinkle plant, to painkillers from willow bark, and even contraceptives from yams! Plants have provided modern medicine with more diverse and important drugs than any other natural source.

As technology and time progressed some of these natural drugs were synthesized and modified to improve or enhance their properties and as a result these natural products were largely supplanted by their synthetic counterparts. Successes in this field tended to overshadow the pharmaceutical industry's roots in natural products and it was generally thought that ultimately all the plant drugs would be obtained from synthetic sources. However, not all drugs can be commercially produced by synthesis and to this day pharmaceutical chemists still draw on plants when they search for new drug molecules. In the United States alone, pharmaceutical products originating from plants still make up some 25 per cent of prescription drugs.

## Biotic Resource

Estimates of the total number of higher plants species, both identified and unidentified, range from 250 000 to 5000 000 species and even to a higher figure of 750 000. However, most botanists would stick to a conservative figure of 250 000. Whatever the true figure is, one important factor is that only a small percentage of these plants have ever received any more than superficial screening. A great many of those screened comprise plants from the temperate and subtropical regions, due to the fact that phytochemical studies and medical advances in the temperate areas, involving temperate and subtropical plants, have taken place for a much longer period than those for the tropical regions.

The tropical rain forest is hailed to be the most biogenetically diverse of all the forested areas of the world. In a sample plot of 1 hectare of tropical rain forest, up to 100 tree species may be found, compared with only about 10-15, rarely up to 35, at the most, in a temperate forest. It covers only about 7 per cent of the earth's land surface, yet it is home to more than half the world's species of flowering plants. Undoubtedly, a vast storehouse of valuable new phytochemicals still awaits discovery. For example, nothing is known about 99 per cent of the flora of Brazil! Who knows what wonder drug still lies in wait within these dense and dark forest walls?

## Role of Plants in Drug Discovery and Development

Plants produce a highly individual range of natural products which vary widely from species to species and are mostly structurally distinct from microbial metabolites. There are essentially four basic ways in which plants contribute to modern medicine. First, plants are sources of direct therapeutic agents. For example, the South American jungle liana *Chondodendron tomentosum* is the main source of d-tubocurarine, a muscle relaxant much used in surgery. Chemists so far have been unable to produce this drug synthetically in a form which has all the attributes of the natural product. Furthermore, there are incidences where, even when chemical synthesis is possible, it is less costly to harvest the drug from its natural sources. One such example is the hypotensive drug reserpine which is still commercially extracted from *Rauwolfia* species.

Plants are also a starting point for the elaboration of more complex semi-synthetic compounds. Among the most important therapeutic agents used in modern medicine today are steroidal drugs such as the corticosteroids, sex hormones, anabolic agents and oral contraceptives. Although these drugs can be obtained from a number of sources, including total synthesis, steroidal sapogenins obtained from plant species, for example diosgenin, which may be obtained from tubers of various species of *Dioscorea*, constitute one of the major raw materials for the partial syntheses of these drugs. At present the use of diosgenin has decreased by half or less due to the widescale use of stigmasterol and sitosterol which are also obtained from plant sources. Nevertheless diosgenin is still being used and Dioscorea species will continue to be an important source for steroidal drugs at least in the developing countries.

Plants are also a source of natural products which serve as models for new, pharmacologically active compounds in the field of drug synthesis. There are several reasons for this. It may be that the plant material is not present in abundance and large scale cultivation is not viable, precluding the direct use of the natural source. Another reason is that in some cases the side effects of a natural product often prevent its use in medicine and can be resolved only by preparation of a synthetic derivative; examples are cocaine, a template for other modern local anaesthetics, and modifications of podophyllotoxin to obtain other antitumour preparations. New and unusual chemical substances found in plants will continue to serve as models for novel synthetic substances and will prove to be increasingly important in the future.

Finally, plants may also act as a natural source of compounds whose side effects are too strong to permit their use as prescription drugs, but which are valuable in research such as in the investigation and characterization of biochemical process and their mechanisms. This is a very important, though obscure, use which assist in drug discovery and development. Many compounds with anticancer properties have been found to be toxic for use as clinical drugs but nevertheless are widely used and have proved to be very helpful in research

## Progress in Plant Drug Research

The 1950s coincided with a number of significant events such as the discovery of reserpine, the start of the investigation of the vinca alkaloids, and the development of refined chromatographic procedures and radioactive tracer techniques for biosynthetic studies. These provided considerable impetus to the investigation of drugs from plants. Apart from significant achievement in the field of drug synthesis, during the 40 years that followed, major advances were also continually being made in other areas related to plant drug research.

Much of the progress achieved in plant drug research today has been due to the analytical instrument and methods developed and employed during the last 40 years. Thin layer chromatography (TLC) and liquid chromatography techniques were widely used and remain of considerable importance to this day. Since such techniques were first developed they have undergone tremendous improvements and innovative modifications enabling better separations of mixtures of plant products. Advances in electronics brought more efficiency and sensitivity to spectroscopic techniques, especially nuclear magnetic resonance spectroscopy (NMR), mass spectrometry (MS) and X-ray crystallography. These analytical instruments are now more sophisticated and readily available; they are favoured as indispensable methods for structural determination. The spectral methods have been combined with chromatographic techniques, such as GC/MS, HPLC/MS and HPLC/NMR which permit the direct identification of separated compounds with remarkable ease.

The lack of simple bioassay procedures has been a continuing source of problems for natural products chemists in determining the physiological activity of plant materials, whether in the form of crude fractions or as purified chemical entities. Previously, fairly elaborate assays were used, for example the rat 'Hippocratic' screen. This was followed by the more successful brine shrimp (*Artemia salina*) toxicity assay, and the potato-disc assay which involves observation of the inhibition of crown-gall tumours induced on potato discs by *Agrobacterium tumefaciens* Conn. These methods were found to be rapid, reliable, inexpensive, and may be conveniently applied in-house by natural products chemists. In recent years major advances in bioassay techniques have taken place, in parallel with automated high-throughput screening technology based on the use of microelectronics, robotics and advanced spectroscopic instrumentation. The advent of modern biotechnology has led to the development of 'mode of action' bioassays including immunoassays capable of detecting picogram quantities of potentially useful compounds. Advances have also been made in areas of molecular and biochemical pharmacology which facilitated the development of assays for compounds which can selectively inhibit, or bind to, enzyme and receptors associated with known physiological events. Such integrated systems can screen thousands of samples daily and efficiently pinpoint those with pharmaceutical utility.

Another area relevant to plant drug research is the production of plant material for an adequate supply of the drug for clinical use. As civilization encroaches on forested areas, collection of plant material from the wild becomes less and less feasible. Drug producing plants do not often lend themselves to cultivation easily and agronomic research of drug producing plants has been somewhat limited because these plants were considered to be of relatively minor economic importance. Resorting to synthetic production of these drugs may not be as easy as it sounds as most often the structural complexity inherent in such natural products demands multi-step syntheses, which, although of distinct academic interest, are rarely of practical utility for large-scale industrial production. A solution to this question of increasing material availability and eliminating dependence on the living plant as the source is production using plant-tissue and cell culture techniques. As well as having great potential commercially, these techniques have been useful in the study of plant biosynthesis and regulation of plant secondary metabolite production. Although there are still limitations to such techniques such as slow growth, expensive media, and the tendency to store desired metabolites in the tissues rather than excrete them into the media, cell suspension cultures seem to be the most appropriate system for the production of secondary products on an economical scale, provided that strategies are developed to shorten fermentation times and increase yields. Currently, certain pharmaceutically important chemicals such as shikonin, digoxin, vinblastine and rosmarinic acid are being successfully produced commercially by cell culture in large bioreactors.

## Some Significant Plant Drugs

Based on computerized information in the NAPRALERT database on natural products, there are currently about 125 clinically useful prescription drugs worldwide, derived from only 95 species of higher plants. A few of the more important drugs will therefore be briefly discussed in the following sections. At least 45 of the 125 drugs listed are derived from about 39 plants, originating in and around the tropical rain forests and almost half of these drug-yielding tropical species are Asian plants.

### Drugs for Heart Diseases

The American foxglove (*Digitalis* species) has been used for medicinal purposes for hundred of years but it was only in the late eighteenth century that it was shown to be effective in the treatment of heart disease. It is the source of the digitalis drugs such as digitalin digoxin, acetyldigitoxin, gitalin, lanatosides A, B, C, etc. These cardiac glycosides encompass compounds which contain a cardenolide linked to one or more glucose-like moieties and have a positive inotropic action on the heart. *D. purpurea* and *D. lanata* are two main sources of the digitalis drugs which are the treatment of choice for arrhythmias and heart failure.

A number of drugs used for cardiovascular disorder are drugs derived from tropical trees. The best example is probably the well-known alkaloids of *Rauwolfia serpentina*, used as antihypertensives and as tranquilizers. *R. serpentina* occurs throughout India, Malaysia and

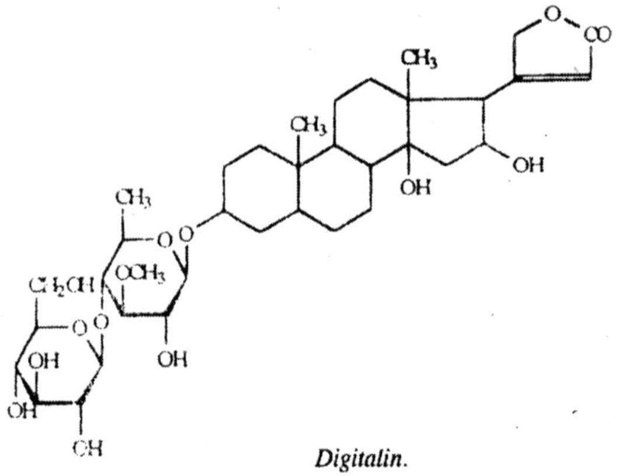

*Digitalin.*

Thailand as small trees that grow wild in the humid forests. It is now cultivated in many tropical countries. Because of its highly toxic nature, use of *R. serpentina* has decreased considerably and a sister species, the African serpent wood *R. vomitoria* Afz., has been exploited much more for the world market.

*Rauwolfia* alkaloids used as drugs include reserpine, rescinnamine, deserpidine, ajmalcine and ajmaline. Reserpine is used in combination with diuretics for controlling mild to moderate hypertension. It depletes peripheral nor-adrenaline stores, resulting in a fall in peripheral resistance and blood pressure as well as bradycardia and CNS

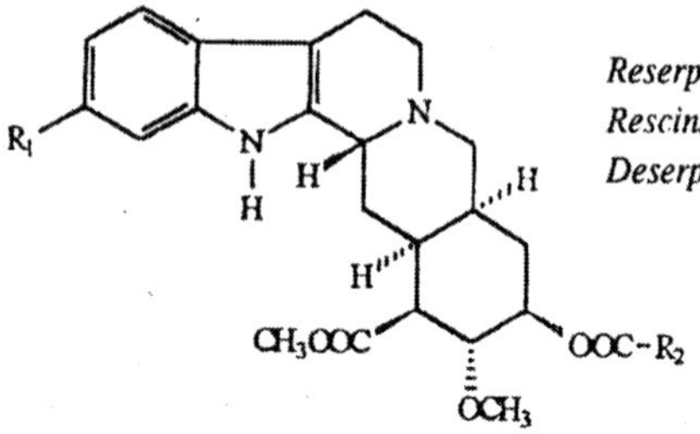

| | $R_1$ | $R_2$ |
|---|---|---|
| *Reserpine* | $OCH_3$ | $C_6H_2(OCH_3)_3$ |
| *Rescinnamine* | $OCH_3$ | $CH{=}CHC_6H_2(OCH_3)_3$ |
| *Deserpidine* | $H$ | $C_6H_2(OCH_3)_3$ |

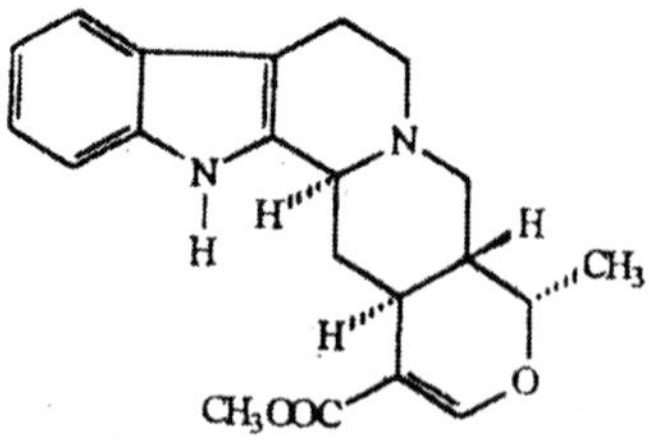

*Ajmalicine*

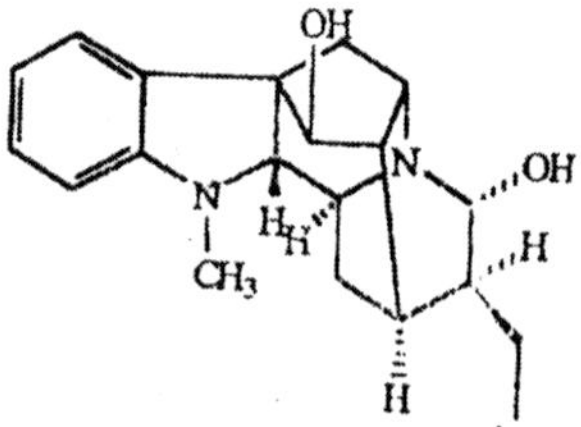

*Ajmaline*

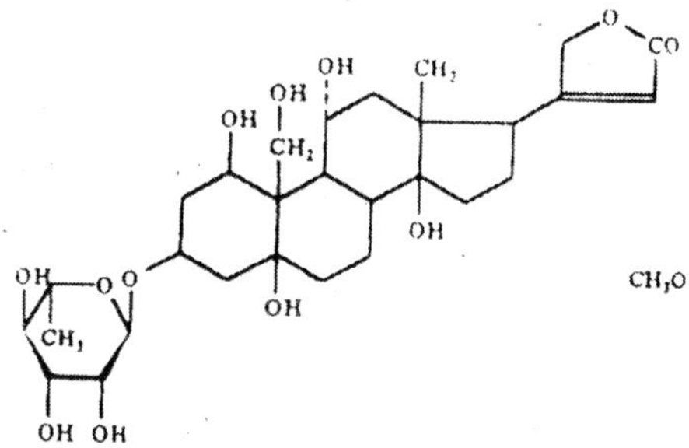

*Ouabain*

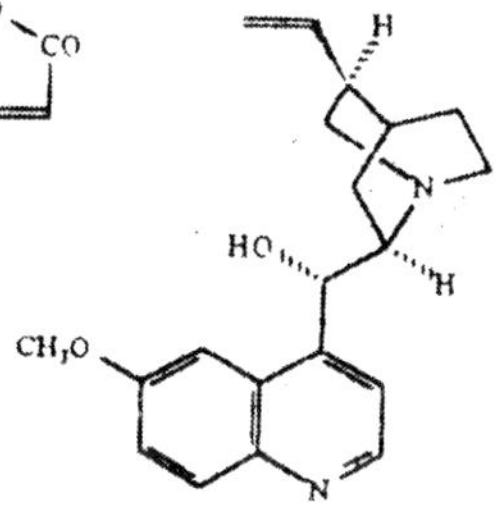

*Quinidine*

depression. It was also formerly used to treat psychotic disorders. Rescinnamine and deserpidine have properties and uses similar to reserpine. Ajmalicine, also known as raubasine, has also been used in conjunction with other agents to treat hypertension and in peripheral and cerebral vascular disorders. Ajmaline has antiarrhythmic activity on the heart muscle and is used clinically as a therapeutic agent in cardiac arrhythmia, as well as being used as an antihypertensive and tranquilizer. This compound is found in very large quantities in *R. vomitoria* and has become much more popular as a hypotensive agent than reserpine. Other drugs for cardiovascular disorders derived from tropical species are the cardiotonic, ouabain and the antiarrhythmic quinoline alkaloid, quinidine. Ouabain is an injectable cardiac glycoside, extracted from the seeds of *Strophanthus gratus*. It has a faster onset of action than the usual digitalis digoxin, hence its preferred use over the latter when rapid benefit is required and in emergency situations. Like digitalis, ouabain is used to treat atrial fribrillation (arrhythmia) with an uncontrolled ventricular rate, atrial flutter, supraventricular tachycardia and acute left ventricular failure. Quinidine is extracted from the bark of *Cinchona ledgeriana* and used as a cardiac depressant or antiarrhythmic. Kawain is a naturally occurring pyrone found in the rhizomes of *Piper mythesticum*, a shrub indigenous to islands of the South Pacific. Currently produced synthetically, it is used as a tranquilizer and to improve well-being in geriatric patients.

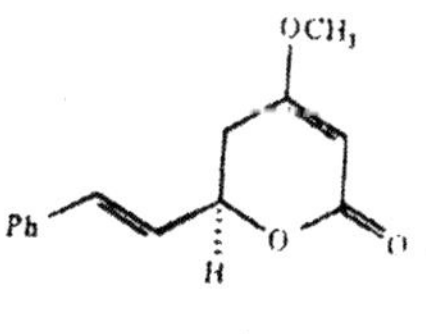

*Kawain*

**Local Anaesthetics**

Cocaine or 2R-methoxycarbonyl-3S-benzoyltropine is found in the leaves and barks of the South America shrub *Erythroxylon coca*. It is one of the major coca alkaloids. The South American native have been known to chew the coca plant to stimulate quick recovery from fatigue. The plant is now cultivated in a number of countries including Peru, Bolivia, Colombia, Indonesia and Sri Lanka.

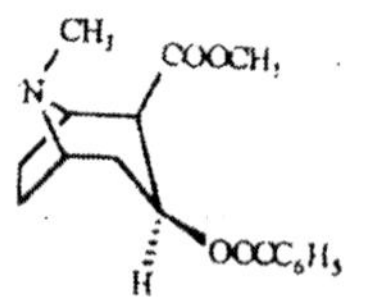

Cocaine is used as a topical local anaesthetic, to relieve pain in cancer patients and to relieve pain from cluster headaches. Its anaesthetic action comes from its reversible membrane stabilizing effect. It is rapidly absorbed after topical administration and has a vasoconstrictive action, thus enhancing its effectiveness as a local anaesthetic. It also functions as a central nervous system stimulant. Cocaine has been found to cause narcosis and its abuse leads to addiction.

*Cocaine*

## Analgesics

Morphine and codeine are the two well-known opioid analgesic drugs. These alkaloids may be extracted from the dried sap obtained by lancing the unripe seed pods of the opium poppy (*Papaver somniferum*) or by solvent extraction of poppy straw. The most important opium-producing countries are India, turkey, Bulgaria, Yugoslavia, USSR, Australia, France and Spain. To date, morphine is still considered to be the drug of choice for the control of acute and chronic pain of malignant origin such as cancer. Its clinical use is dependent on its interaction with opioid receptors in the brain, spinal cord and gut. It is also employed for treatment of typhoid fever, traumatic shocks and, in combination with atropine sulphate, for relieving renal and intestinal colic and coronary thrombosis. Codeine is less potent than morphine in its pain relief capacity; it is thus used for the control of mild to moderate pain. Both morphine and codeine possess antitussive properties and have been used as cough suppressants. Codeine is however more acceptable for such uses because morphine tends to increase the incidence of post-operative chest complications by the suppression of a productive cough.

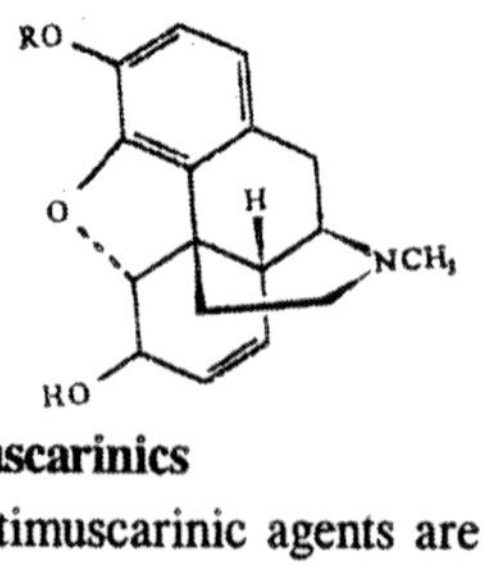

*Morphine: R = H*

*Codeine: R = CH*

## Antimuscarinics

Antimuscarinic agents are competitive inhibitors of the actions of acetylcholine at the muscarinic receptors of autonomic effector sites innervated by parasympathetic nerves. Plant drugs included in this class are the belladonna alkaloids atropine, hyosycamine and hyoscine. Atropine or DL-hyosycamine is the chief alkaloid extracted from the deadly nightshade, *Atropa belladonna*, *Datura stramonium* and several other Solanaceae plants. A belladonna is indigenous to Western Europe and cultivated in England, Germany, USSR, USA and India.

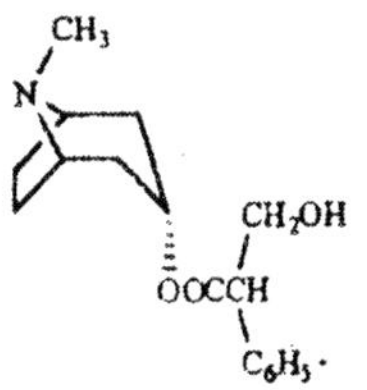

*Alkaloids atropine* 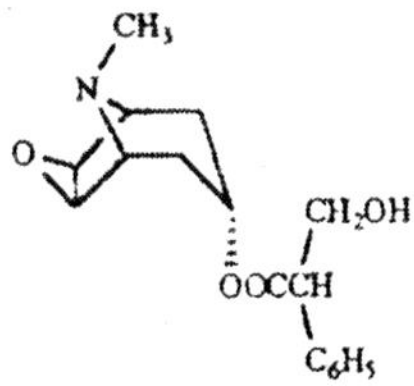 *Hyosycamine and hyoscine*

Atropine may be prepared synthetically or by racemization of the naturally occurring L-hyosycamine. Atropine is the prototype and best known antimuscarinic agent although many of its uses are now superseded by other semi-synthetic antimuscarinic drugs. It is used primarily for the treatment of stomach spasm and also applied topically to the eye to produce dilatation during ophthalmic examinations. Hyoscine, also known as scopolamine, is a closely related ester of atropine. Although it can be produced synthetically, it is usually obtained by extraction from various members of the Solanaceae, including the well-known herb, *Datura metel*. The drug also has an anticholinergic effect but in contrast to atropine it also has a CNS depressant effect, hence its use as a sedative and to treat motion sickness.

## Miotics

Pilocarpine and physostigmine are two well-known cholinergic drugs used as miotics in the treatment of glaucoma. Pilocarpine, an alkaloid obtained from the leaves of *Pilocarpus jaborandi*, is a direct-acting muscarinic parasympathomimetic agonist. Physostigmine, an alkaloidal constituent of the calabar bean of the woody vine, *Physostigma*

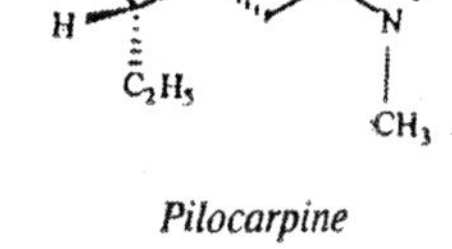

*Pilocarpine* *Physostigmine*

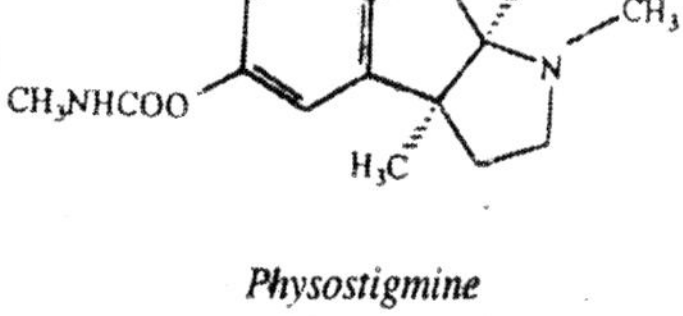

*venemosum* is an indirect acting parasympathomimetic agent. Used as the hydrochloride or nitrate, pilocarpine is the first choice when miotics are required to reduce intraocular pressure in the treatment of open-angle glaucoma. This is due to the fact that pilocarpine generally provides good control of intraocular pressure with relatively few adverse effects. Physostigmine is not as well tolerated as pilocarpine, hence is rarely used for long-term therapy. However physostigmine has been used for more than a hundred years as an antidote for atropine overdose and more recently in poisoning with tricyclic/tetracyclic antidepressant drugs.

## Muscle Relaxants

*d*-Tubocurarine is perhaps the most well-known of the natural skeletal muscle relaxants. The alkaloid is the active principle in 'tubocurate', the arrow poison used by the South American Indians in the Amazon-Orinoco basin. The compound can be obtained from extracts of the stems and bark of the liana *Chondodendron tomentosum* and several other species of this genus. Available pharmaceutically as the chloride, it is competitive neuromuscular blocker, primarily used intravenously to produce skeletal muscle relaxation during surgical procedures. It acts by competing with acetylcholine for receptors on the motor end-plate to produce neuromuscular blockade, seen as flaccid paralysis. A patient is given the drug to reduce the amount of general anaesthetic required to achieve total muscle relaxation, although the resultant paralysis of the respiratory muscle means that artificial ventilation of the patient is necessary.

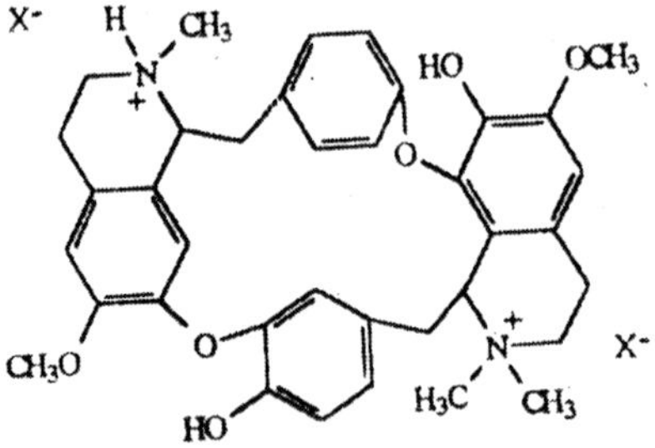

*d-Tubocurarine*

Papaverine is a smooth muscle relaxant and vasodilator. The alkaloid is also extractable from the opium poppy, *Papaver somniferum*, but unlike the other opium derivatives, is not habit forming. Its use has been largely replaced by drugs with more specific actions such as

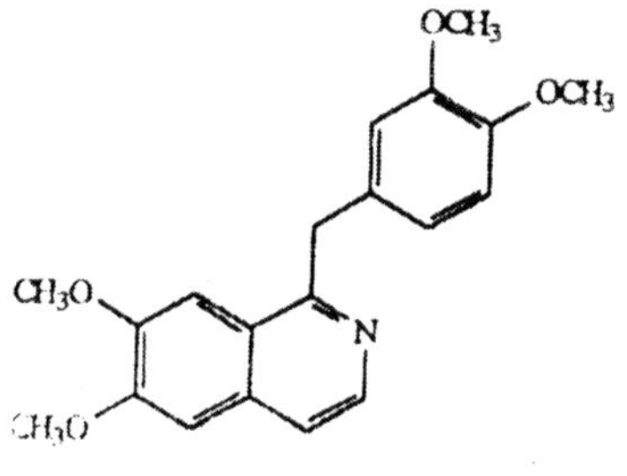

*Papaverine*

α-adrenergic blockers and calcium slow channel antagonists but it may still have a place in the treatment of vascular spasms.

**Bronchodilators**

The ephedra alkaloids obtained from the Chinese plant 'Ma Huang' (*Ephedra sinica*), ephedrine, pseudoephedrine and norseudoephedrine, are sympathomimetic agents with direct and indirect effects on adrenergic receptors. These drugs are now produced synthetically. Ephedrine is used for the treatment of nasal congestion and as a bronchodilator for treating the symptoms of asthma. Pseudoephedrine, its naturally occurring stereoisomer, and nor-pseudoephedrine are also used for the same purpose but possess less potent pharmacological properties. The two drugs aer widely used as constituents in over-the-counter remedies for treating symptoms of the common cold. They have also been found to show significant CNS excitatory and pressor effects which makes them undesirable for use in hypertensive patients. Ephedrine is no longer a drug of choice in the treatment of asthma as more selective drugs with less cardiac and CNS stimulation effect are available.

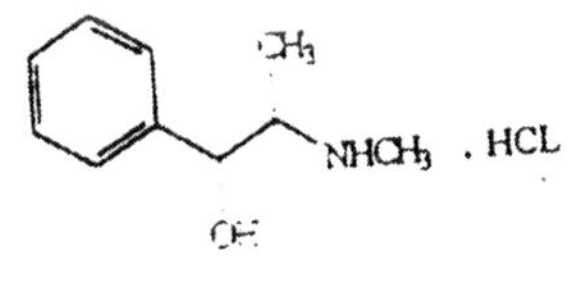

*Ephedrine*

**Antineoplastic Agents**

Cancer is a serious life-threatening disease for modern society today. Among the most prevalent forms of cancer are breast, colorectal, lung, ovarian, prostate and uterine. Many different structural classes of plant secondary metabolites have been found to be cytotoxic but only a few are used clinically.

The most frequently used anticancer drugs are the dimeric indole alkaloids, vinblastine and vincristine, which have been available since the 1960s. First discovered from the Madagascar periwinkle *Vinca rosea* (*Catharanthus roseus*), they have become the two most important

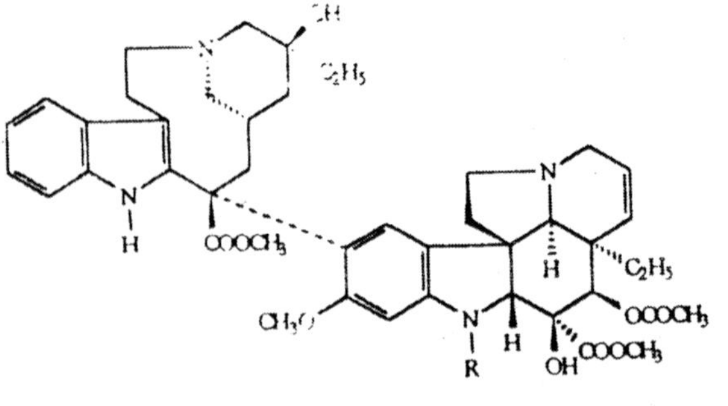

*Vinblastine: R = CH$_3$; Vincristine: R = CHO*

clinically useful anticancer agents from any plant source. The durgs are still extracted from natural sources. Most of the *C. ròseus* used for production of the two drugs is grown under cultivation in India, Madagascar, Israel and the USA.

Vinblastine and vincristine are used, either singly or in combination therapy, in the management of malignant diseases, particulary lymphomas and sarcomas. Hodgkin's disease and the leukaemias. These vinca alkaloids exert their biological effects by binding specifically with the protein tubulin, inhibiting its assembly into microtubules with resultant dissolution of the mitotic spindle which eventually leads to cell death. Although not a first-line drug, vinblastine and vincristine have also been used in combination with other anticancer drugs for breast cancer.

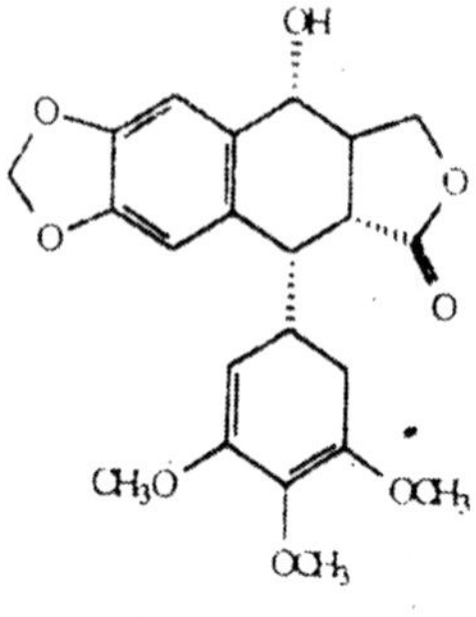

Podophyllotoxin is the active principle in podophyllin, the alcoholic extract of the dried rhizomes and root of the North American May apple, *Podophyllum peltatum*. The drug has applications in dermatology where it is an effective therapy for anogenital warts, and possibly nasal papillomas as well as psoriasis.

*Podophyllotoxin*

Like the vinca alkaloids, podophyllotoxin is also a microtubule inhibitor. This compound and its congeners have however been found to attack both normal and cancerous cells. The toxic side-effects of these lignans have limited applications as drugs in cancer chemotherapy, except for etoposide and teniposide which are semisynthetic derivatives of podophyllotoxin. These two antitumour agents do not exhibit any effect on intracellular microtubules but induce breaks in single and double stranded DNA, through their interactions with topoisomerase II which is a critical enzyme in DAN replication.

**Antiprotozoals**

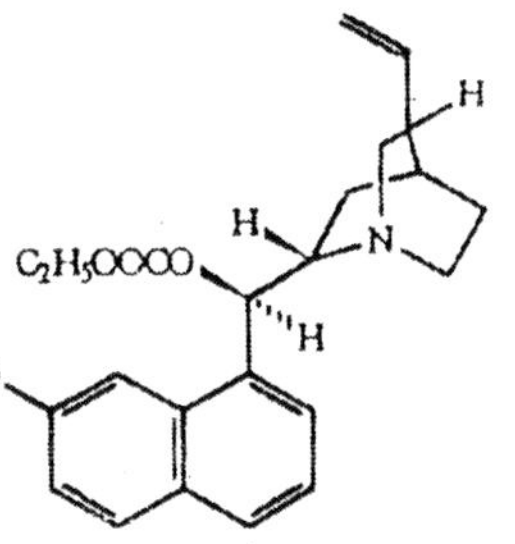

Protozoal infections are responsible for a number of major diseases including malaria, amoebiasis, leishmaniasis, giardiasis and trypanosomiasis. These diseases affect millions of people worldwide, both in the developing and developed part of the world. The AIDS epidemic has resulted in an increase in infections due to *Cryptosporidium parvum* which causes severe diarrhoea and *Pneumocystis carinii* which results in pneumonia. In the last century there have been two major antiprotozoal drugs obtained from higher plants: the antimalarial drug, quinine, and the amoebicidal drug, emetine.

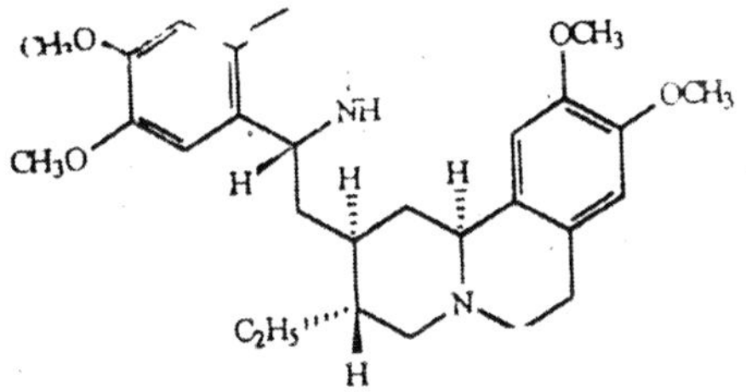

*Quinine*

*Emetine*

Quinine, a quinolinemethanol, was used in the treatment of malaria long before the malaria parasite was even discovered. It is extracted from the bark of various species of *Cinchona*, in particular *C. ledgeriana*, where it occurs together with its diextrorotaory stereoisomer quinidine and twop othe main alkaloids, cinchonidine and cinchonine. The main *Cinchona* producing countries are Indonesia, Zaire, Tanzania, Kenya, Rwanda, Sri Lanka, Bolivia, Colombia, Costa Rica and Indian.

Quinine is used primarily for the treatment of severe and complicated *Plasmodium falciparum* induced case of malaria, especially when the cases are resistant to synthetic chemotherapy. The chemical structure of quinine has also served as a template molecule for the design and development of several other antimalarial drugs including chloroquine. At one time quinine was superseded by synthetic antimalarials which have fewer side-effects. However resistance of *P. falciparum* to antimalarial chemotherapy favours the continued use of quinine.

The other three *Cinchona* alkaloids have also been shown to be effective in the treatment of falciparum malaria. Only quinidine however, which is actually superior to quinine in its antimalarial effect but more likely to cause cardiac toxicity and hypersensitivity, has been recommended for oral or parenteral use in the event of quinine unavailability.

Emetine is a long recognized effective therapy for invasive amoebiasis, widely employed in developing countries for treating amoebic dysentery. Used in the form of the hydrochloride salt, the alkaloid is obtained from ipecac which is the dried root of the Brazilian plant *Cephaelis ipecahuanha*. It may also be semi-synthetically obtained by methylation of *cephaeline*, the othe rmain constituent of ipecac. Ipecac root products (particularly ipecac syrup) are widely used as emetics in cass of poisoning. Emetine has a direct lethal action on the protozoan *Entamoeba histolytica* in tissues, including bower, invaded by the organisms. It has no effect however an amoebae confined to the lumen of the bowel. High doses of emetine are toxic to humans. It has severe adverse effects, exhibiting toxicity to heart, liver, kidneys and gastrointestinal tract. It is currently superseded by the synthetic metronidazole and its use is restricted to initial treatment of severe amoebic abcesses. Another plant drug, glaucarbulin, which is a quassinoid-type degraded triterpene and is extracted form the American Simarouba glauca, has also been used in the oral treatment of amoebic dysentery.

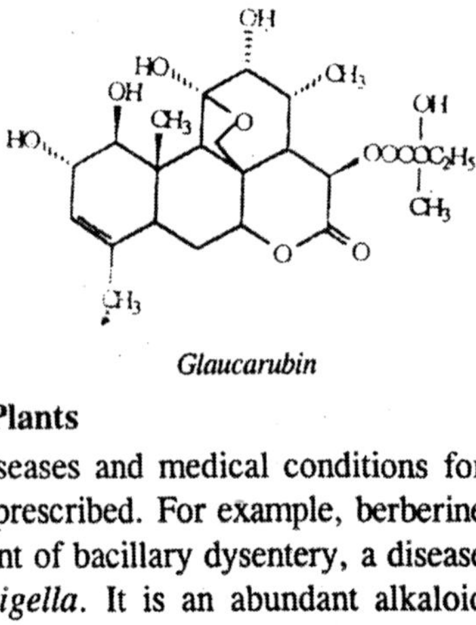

*Glaucarubin*

**Other Miscellaneous Drugs from Plants**

There are a myriad of other diseases and medical conditions for which drugs derived from plants are prescribed. For example, berberine is a widely used drug for the treatment of bacillary dysentery, a disease caused by bacteria of the genus *Shigella*. It is an abundant alkaloid extractable from *Berberis vulgaris* and many other plant sources of the families Annonaceae, Ranunculaceae (mostly herbaceous), Berberidaceae, Menispermaceae and Papaveraceae. It has good antibacterial activity and is poorly absorbed form the gastrointestinal tract when administered orally. Anisodamine and anisodine, the atropine-like alkaloids

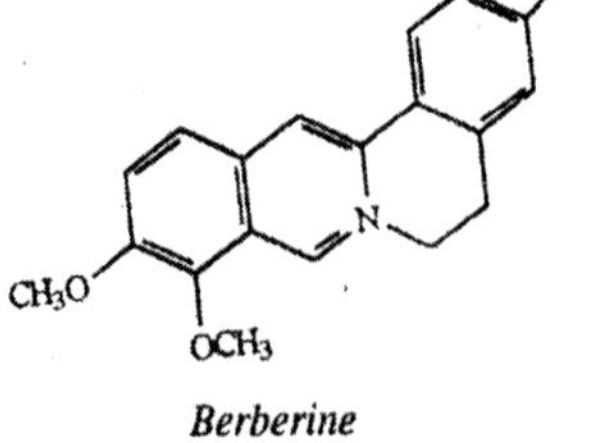

*Berberine*

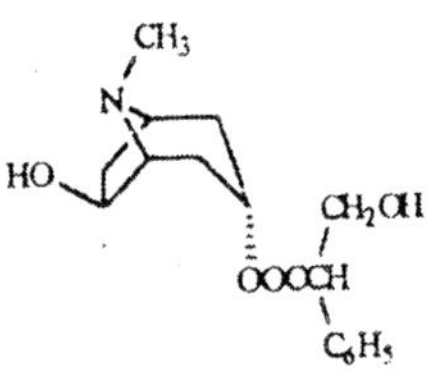

*Anisodamine* 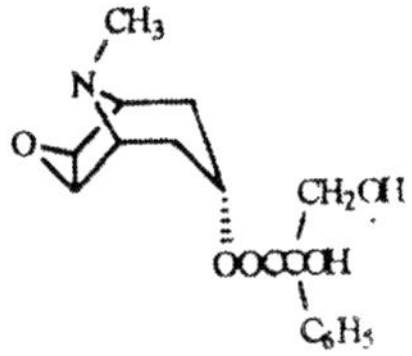 *Anisodine*

isolated form the Chinese painkiller plant, *Anisodus tanguticus*, are anticholinergic agents. The quinolizidine sparteine from *Cytisus scoparius* is employed as an oxytoxic in childbirth to initiate uterine contractions and to decrease post-partum haemorrhage. Vasicine, a quinazoline alkaloid form the Indian species *Adhatoda vasica*, has a similar use. The sennosides, derived from *Cassia angustifolia* and *C. acutifolia*, are used as laxatives. Sanguiñarine from *Sanguinaria canadensis*, is used in dental preparations due to its ability to prevent dental plaque formation.

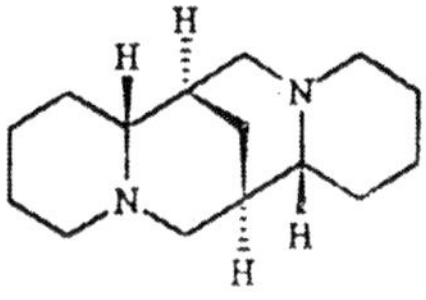

*Quinolizidine sparteine* 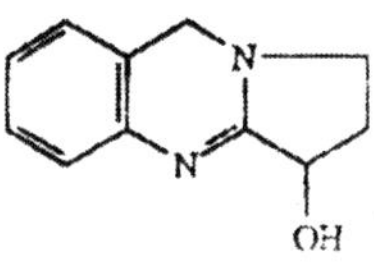 *Vasicine*

## RECENTLY DISCOVERED PLANT-BASED DRUGS

Since the 1950s, spurred by the discoveries of the rauwolfia and vinca alkaloids, numerous plant species have been studied for their chemical constituents and biological properties. In 1985 NAPRALERT listed almost 200 000 compounds which have been identified from natural sources, 70 per cent of which were plants. This number may well have increased significantly since then. Success in the development

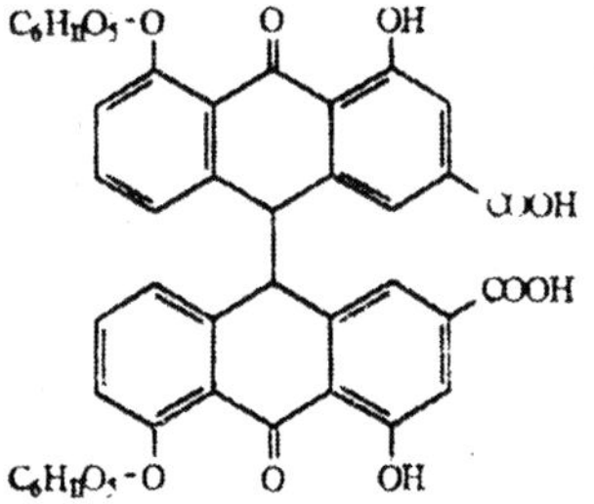

*Sennosides* 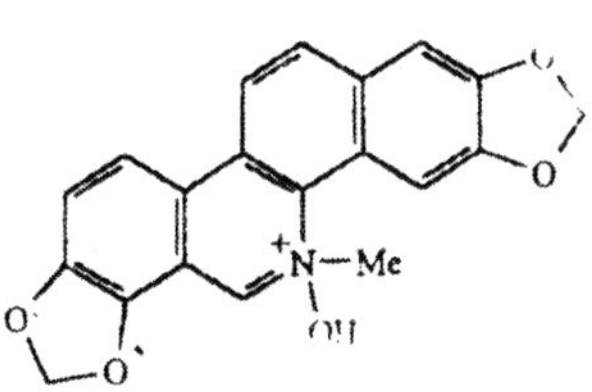 *Sanguinrine*

of new, commercially viable drugs has however been modest. Since the last major breakthrough of vinblastine and vincristine, the only othe drug that the world has heralded with equal excitement has been taxol, which is also an antitumour agent. The discovery of the antimalarial, artemisinin, was also of particular importance, especially for tropical countries. Nevertheless many other potential drugs are still at their early stages of development and clinical trials and the outlook for the future of drugs from plants remains optimistic.

**Taxol and Camptothecins**

Current advances in cancer chemotherapy are due to the discovery of two classes of natural products, the taxoids and the camptothecins. Taxol or paclitaxel was first discovered in 1971, from the bark of the northwest pacific yew tree, *Taxus brevifolia* Nutt. Since its discovery, structure elucidation and biological activity more than 20 years ago, taxol has finally been accepted as an anticancer agent, culminating in its recent FDA approval for use in the treatment of breast and ovarian cancers, two of the most difficult forms of cancer to treat. Taxol's mode of action is by targeting microtubule formation. Unlike other antimicrotubule agents, including the vinca alkaloids, which block microtubule production, however, taxol promotes tubulin polymerization and stabilizes microtubules against depolymerization. Much remains to be learned about the clinical use of taxol, such as its activity towards other forms of cancer dose-response relationships, effectiveness in combination with other drugs in combination therapy, etc. Nevertheless, based on just the ovarian and breast cancer utilization, worldwide need of taxol has been estimated to be hundreds of kilograms per year. The yield of taxol from its rare and slow-growing natural source is very low, less than 0.02 per cent dry weight. If the current trend of promising activities against other cancer forms, countinues larger supplies and alternative sources of taxol will be needed and

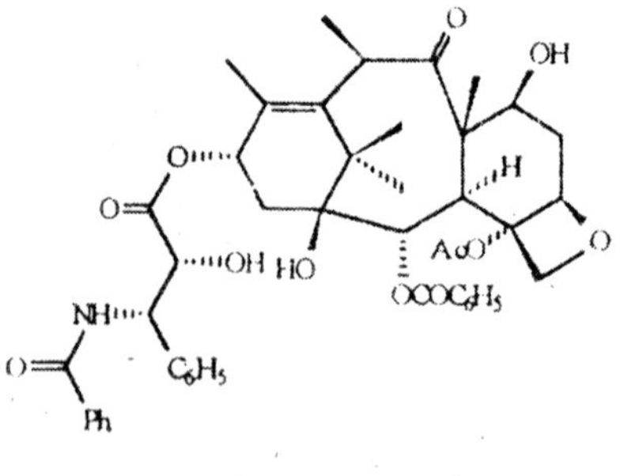

*Taxol or Paclitaxel*

many mature trees would have to be felled in order to supply sufficient drug for clinical use. Although the semi-synthesis of the compound has been achieved, it will be probably still not answer the supply problem. Efforts at solving this problem through tissue culture and microbial fermentation are currently underway and show great promise of success.

Camptothecins, the second group of natural antineoplastic compounds, are derivatives of the isoquinoline alkaloid camptothecin, first isolated from the Chinese tree *Camptotheca acuminata*. The alkaloid has been subjected to limited clinical trials but its effects have not paralleled those seen in animal studies. The compound was found to have a serious side-effect, causing bleeding in the bladder and kidneys. Better results have however been observed in clinical trials of the relatively new camptothecian derivatives, topotecan and irinotecan. Whereas several families of topoisomerase II inhibitors, such as anthracyclines, epipodophyllotoxins, acridines, and ellipticines, are known and widely used in clinical practice, the camptothecins are the unique representative of selective topoisomerase I inhibitors with clinical applications.

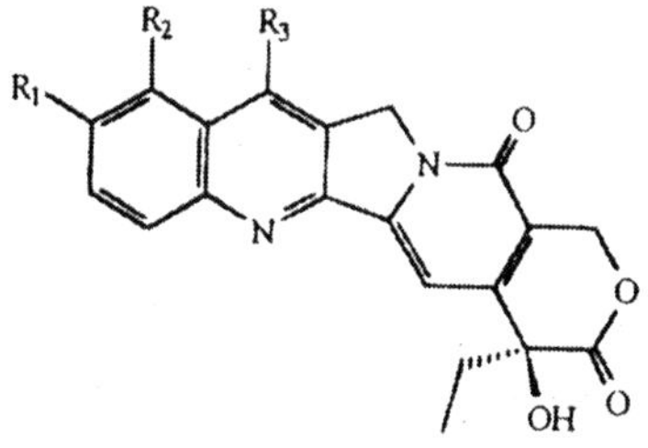

| | $R_1$ | $R_2$ | $R_3$ |
|---|---|---|---|
| *Camptothecin:* | *H* | *H* | *H* |
| *Topotecan:* | *H* | $CH_2NH(CH_3)_2$ | *OH* |
| *Irinotecan:* | $CH_2CH_3$ | *H* | |

## Artemisinin

Until the recent discovery of artemisinin from the Chinese antimalarial plant, *Artemisia annua* (Qinghao), there had, for many years, been no further antiprotozoal drugs from higher plants. Also known as quinghaosu, this is an unusual endoperoxide sesquiterpenee lactone. Artemisinin has been found to be an effective antimalarial drug against both chloroquinine-resistant and chloroquinine-sensitive strains of *Plasmodium falciparum* as well as as against cerebral malaria.

*Artemisinin:* $R = O$

*Dihydroartemisinin:* $R = OH$

$R = -OMe$

$R = -OEt$

$R = -OCHOCH_2CH_2COONa$

Artemisinin is currently in clinical use and in some parts of south-east Asia it seems to be the only effective drug against infections from multidrug resistant *P. falciparum*.

The significant biological activity, novel chemical structure and its low yield from natural sources have promoted efforts directed at its synthesis. In the past few years, a number of derivatives have been developed and found to be clinically active against malaria, including multi-drug resistant *P. falciparum*. For example, reduction of the lactone carbonyl of artemisinin yields dihydroartemisinin from which the semi-synthetic products have been prepared. The World Health Organization and Rhone-Poulence Rorer have in fact collaborated in the development of artemeter. Registration of these products has been approved in six African countries and applications field in another eight countries. The current interest in artemisinin is considerable and it is possible that other antimalarial drugs based on its structure will be developed in the future, in a similar way to quinine being used as a template molecule for the development of chloroquine and mefloquine.

## Summary and Conclusions

It is apparent that the plant kingdom is a rich source of biologically active natural products and no doubt it will continue to serve mankind in the future just as it has done since the dawn of history. This is likely in view of the fact that only a small proportion of plants, especially those of the tropical rain forests, have been thoroughly investigated for their medicinal potential. Plants are useful in their crude or advanced forms as drugs, and biologically active compounds from plants can serve as templates for the synthesis of modern drugs. Method of plant drug research have improved tremendously over the years, and much more has been learned about the basics of plant metabolism, plant analysis and even plant production. Looking ahead to the future, there are still many diseases such as AIDS, certain cancers, Parkinson's disease, muscular dystrophy, and cystic fibrosis which require improved and/or satisfactory cures. It is hoped that significant new plant drugs and new methods of producing them will continue to be discovered and developed.

# 11

# PACKAGING FOOD

From the standpoint of problems involved in the use of packages, the principal distinctions between rigid and nonrigid containers lie in (1) storage of the empty containers and (2) character of filling and sealing machinery required. A container which permits no gas or vapour to pass through its seams, its top, its bottom, or its walls is a hermetic container. Containers of some types, ordinarily considered to be hermetic, do not meet the requirements of this definition. Sometimes, a seam of a container functions as a filter, permitting gas to pass through but purifying the gas during its passage so that no solid entities, even those as small as microorganisms, pass through with the gas. This condition exists when a paper gasket is used in a rolled seam of a metal container unless the pores of the paper have been filled with a substance that is nonpermeable to gas. It exists also when certain types of rubber gaskets are used on glass containers.

The supreme test of a hermatic container comes when there is a substantial pressure differential between the interior and the exterior of the container. Such a pressure differential tends to force gas through any permealbe material which separates the inside gas from the outside gas. No nonrigid container for food is classed as a practical hermetic container because no material, ordinarily considered as flexible material, except thin sheet metal, is absolutely impermeable to gas. No way, except soldering, is known to produce a reliable hermetic seam in flexible metal and, even when such a seam is produced, flexing of the metal in use is likely to produce perforations. It is desirable that a rigid container be capable of retaining its shape, in its essential characteristics, under the stress of either internal vacuum or internal pressure. Practically all sterilized foods are held under vacuum, as are also some nonsterile dry foods, such as coffee. Sometimes, air is replaced by an inert gas, of which the most commonly used is nitrogen. This practice most often occurs with powdered products, for example, dry whole milk, in which protection from the effects of oxygen is desired and for which a container is used that lacks the mechanical strength necessary to withstand the physical stress of vacuum.

Some primary food containers are specially constructed to make them capable of withstanding internal pressure exceeding atmospheric pressure by from one to five atmospheres. These containers are used for such products as carbonated beverages and the "whipped-cream" type of products. Foods that have liqueform characteristics contain either water or edible oil in considerable quantity. They may be called moist foods. Moist food products are perishable from action by microorgainisms, from action by enzymes, and from action of chemical nature not associated with microorganisms or enzymes. Because of their liqueform nature, these foods are packaged in materials that are either moisture proof, oil proof, or highly resistant to penetration by water or oil. Metal, glass, plastic, to fibre, coated with either wax or synthetic plastic, are commonly used for containers of moist foods. Because of their perishable nature, these foods are packaged in hermetically sealed containers if they are to be held for a substantial length of time; if the required holding period is so short that spoilage can be prevented even though the causative agents are in contact with the food, non-hermetic containers are used.

Foods that contain insufficient water or oil to give them a moist appearance are dry products. From a packaging standpoint, the category of dry product embraces particulate products which behave as dry products behave in all respects except that, within the particles, they are moist and subject to spoilage like moist products. Examples are fresh fruits and vegetables protected by skins, These, as well as products which are dry throughout, such as cereal products are packaged in baskets, boxes, and bags; the bags are made of fabric, paper, or plastic.

The first man-made containers for food, composed of either wood or stone, served the purpose for dry food products, such as cereals, as well as any open top containers used today. To wet products, these containers offered no protection against spoilage inducing factors, such as microorganisms and oxygen. A primitive means of supplying some protection from microorganisms was with the use of a layer of fat or of wax, which still serves as a domestic expedient. Bladders and parchment paper, either as containers or covers of containers, represented a bit of progress in providing protection for food; however, only when containers were produced which were capable of being cooked could it be said that progress worthy of the name had been achieved towards protection of food from the action of spoilage producing elements.

A tight closure on a food container, however, while essential for accomplishing the preservation of food, which is susceptible to spoilage by microorganisms, is of practically no value unless its use is accompanied by processing of the food to destroy viable microorganisms present therein—the process which resulted from the work of the brilliant French scientist, Nicolas Francois Appert, at the advent of the 19th century. Appert invented the process of sterilization of food by heat . He found the problem of producing a hermetic seal with the bottles and corks that were available to him to be a momentous one. Bitting abstracted Appert's description of his work with containers as follows: "Bottles were used as they were 'most impermeable to air" but as the ordinary ones had openings too small and were otherwise too weak to resist the treatment in corking and heating, he had special bottles made, some having openings of more than four inches. He made the stoppers from 3, 4, and 5 layers of selected cork with the pores running norizontally so as to hinder access of air."

A demonstration of his method in England by Appert in 1814 led to the designing of tin containers by the English. Later Appert used tin cans in his business, manufacturing them in his own factory because he could not get them made satisfactorily elsewhere. He formed the cans of wrought iron, presumably by drawing, then tin plated them. He made round, oval, and rectangular cans—the latter in order to conserve space aboard ship. Because of the extra care and expense involved in both making and closing rectangular cans, however, Appert discouraged their use.

The first English patent on preserving food "in bottles or other vessels of glass, pottery, tin, or other metals or fit materials" was issued to Peter Durand in 1810. In 1824, Appert demonstrated that he had successfully preserved beef in containers holding more than 10 kg each and he received a reward for having done this. One of his containers held 17 kg. The use of tin cans of increased physical strength facilitated heat sterilization procedures at temperatures above 212° F.

Such procedures entered a period of more rapid development than previously; using autoclaves, salt baths, and retorts. The advent of the tin can has been called the first great stride in the food preservation business, as it opened a large field for work in the preservation of meat, vegetables, and fish.

It is believed that commercial food preservation in glass jars was begun in United States by Wm. Underwood in Boston, Mass, in 1821. On January 19, 1825, a United States patent was issued to Thomas Kensett and Ezra Daggett on a process for preserving food in "canisters of tin." By 1840, it is said that glass jars had been largely displaced in commercial canning by the metal container because the problems involved in sterilizing and handling tin cans were less numerous and less complex than those associated with glass containers.

The use of metal containers increased rapidly in Europe and tne United States during the latter half of the 19th century. All seams of tin plate cans soldered. The bottom end of the cans, when made as a separate piece and attached to the wall, or body, of the can, was itself a shallow container, hand -formed in the early styles and mechanically drawn in the later cans. The wall of this shallow container, which constituted the bottom end of the can, was from 1/8 to 1/4 inch high, perpendicular to the plane of the bottom, and of the proper outline to make it fit closely to the outside surface of the wall of the can. Solder was sweated onto the overlap surface of the can wall and the wall, or, as it is sometimes called, the flange or, the skirt of the bottom end. The top end was soldered to the wall of the can in the same manners as the bottom. This also was part of the operation of manufacturing the can. The top end had a round hole in the centre, varying in diameter from 3/32 to about 2¾ inches. The can was filled through this hole, which was then closed and sealed with a disc (cap) and solder, or in the case of the smallest hole, with solder alone.

**Sanitary Cans**

Since the advent of the 20th century, the soldered seam cans, called hole-and-cap cans, have been replaced for heat sterilized foods by cans with rolled seam ends, called sanitary cans; except in the evaporated milk industry, where because of lower cost, soldered seam cans are still used. With the use of a new technique for attaching the ends to cans came a rapid increase in the rate at which cans could be manufactured and in the rate at which they could be filled and sealed. Today, a single machine manufactures food cans of shelf size at the rate of 450 cans per minute and cans of the smaller shelf sizes are filled and sealed in a single production line at rates that "push" 800 per minute. These developments brought a reduction of cost of canned foods, which, together with product quality improvement through the application of scientific method in the production of raw materials and packing operations, has brought a vast increase in per capita consumption of such foods. Against a consumption in the United States of 1,500 million cans, or about 20 cans per person in 1900, the consumption has risen to 23 billion, or about 145 cans per person today. More than half of that increase has taken place since 1935.

Not all of these cans are used for foods which, in the conventional sense, are known as canned foods, that is, foods, that are processed for sterilization. In recent years, larger and larger quantities of frozen foods are being packed in cans; most prominent among these are fruit concentrates. Dry products, such as coffee, powdered milk and pop-corn, also utilize a considerable number of cans. These products are packed in hermetically sealed containers because the preservation of quality in these foods depends on maintaining control of the composition of the atmosphere surrounding the products during storage.

A comparatively small percentage of the frozen foods, particularly of the noncitrus fruits, is put into containers suited for retail distribution and, of that part which is sold at retail, except for citrus concentrates, by far the greatest portion is packed in fibre cartons, mostly folding boxes. The phenomenal expansion in the use of cans for frozen citrus concentrate makes it desirable to present statistics on those cans separately from other frozen foods containers. The numbers given are slightly larger than the actual numbers of 6-ounce cans used because the figures represent the quantity of material packed and not the number of cans used. Since a small part of the frozen citrus concentrate is marketed in larger than 6-ounce cans the actual numbers of cans used are slightly smaller than the numbers given.

For a number of years makers have been trying to perfect a paper container for soda pop and spokesmen for the manufacturers say that when this task is mastered, the industry will "have a look at the beer market." If this purpose is accomplished, in view of the momentous problem involved in adapting a paper container to soft carbonated drinks, the paper container industry will richly deserve to "have a look" at numerous fields in which it is not currently active.

Other food products requiring cans capable of holding pressure of at least 100 psi gauge are now being packed at the rate of about 50 million cans per year-mostly cans of a 12-ounce capacity. These products are largely dairy product formulations containing cream, packed with a soluble gas, such as nitrous oxide, under pressure. They are discharged from the cans as whipped cream or a simulated whipped cream through a pressure-tight valve. The number of cans used for products of this type is increasing at a high rate.

**Glass Containers**

The use of glass containers expanded in volume greatly during World War II, when, because of shortages of metals, they were used for many food products for which metal containers are customarily preferred. With the end of restrictions on the use of metals, a decline in the volume of packers' glassware began in 1947 and continued through 1949, after which the number of glass containers used for all types of food began an upward trend which continued through 1953, during which year the peak volume of World War II was exceeded.

**Table 11.1. End use of packers glass food containers**

| *Product type* | *Percentage on basis of number of containers used* | | | |
|---|---|---|---|---|
| | 1956 | 1967 | 1988 | 1996 |
| Racked beans | 5.91 | 1.55 | 1.70 | 1.82 |
| Catsup, chilli and tomato juice | 5.99 | 9.95 | 7.95 | 5.21 |
| Other vegetables and vegetable juices | 1.66 | 1.81 | 2.20 | 2.69 |
| Cherries, maraschino | 53 | .63 | .88 | 1.20 |
| Fruit juices | 1.76 | 2.27 | 2.44 | 2.34 |
| Fruits | 1.48 | .95 | .74 | .92 |
| Baby foods | 9.89 | 12.40 | 17.51 | 16.93 |
| Meats and fish products | 151 | 2.36 | 2.04 | 1.55 |
| Salad dressings, mayonnaise, and edible oils | 9.46 | 13.20 | 14.92 | 16.64 |
| Pickles and relishes | 5.91 | 12.40 | 12.92 | 12.95 |
| Olives | 1.63 | 1.46 | 1.37 | 1.47 |
| Preserves, jams, jellies, and fruit butter | 9.70 | 13.82 | 13.17 | 11.67 |
| Peanut butter | 5.53 | 3.04 | 3.27 | 3.58 |
| Sirups | 13.88 | 8.62 | 8.77 | 9.57 |
| Honey | 1.66 | 2.15 | 2.09 | 2.47 |
| Flavouring extracts | — | .72 | .74 | .73 |
| Spices and seasoning | 30 | .34 | .33 | .30 |
| Mustard | 1.84 | 2.44 | 2.42 | 2.34 |
| Horse radish | 22 | .26 | .25 | 22 |
| Sauces and meat gravy | 1.28 | .63 | .74 | .73 |
| Vinegar | 2.38 | 1.81 | 1.95 | 2.06 |
| Miscellaneous | 17.48 | 7.19 | 1.60 | 2.61 |
| | 100.00 | 100.00 | 100.00 | 100.00 |

**Paper Containers**

Of even greater importance as food containers than metal cans and glass containers are folding boxes and paper milk containers. A large percentage of all types listed, except collapsible tubes and plastic bottles, is used for foods. Paper, particularly paperboard, is used so extensively in the retail packaging of commodities that the volume of paper business is considered by many to be a reliable gauge of the condition of business in general.

According to the best available information, 36 per cent of folding boxes is used for food-many of these, of course, for frozen foods. Approximately 1 per cent of folding boxes is used for beverages-as secondary containers. Production of paper was approximately 9 per cent greater and production of paperboard approximately 7 per cent greater.

**Requirements of a Container**

The qualities of a container must be carefully evaluated during the process of developing or selecting a container for a specific use. The diversity of these qualities is at a maximum in consumer containers because every retail primary container is in contact with the product on its inside surfaces and must meet exacting requirements in this as well as in many other respects A detailed description of the problems involved in producing all types of food containers, even if confined to retail containers is beyond the possibilities of a discussion of the length of this chapter. The following pages, therefore, will be devoted to treatment of retail containers from a functional standpoint, base upon an example of a proposed analysis of a specific type of container for appraisal purposes leading to a container development project. The type of container chosen for this example is the paper milk container; however, as each functional characteristic enters into consideration, analogous problem pertaining to this characteristic in various types of container will be treated.

As a foundation on which to base an appraisal of a type a container, two series of factors will be used; the first series wig be designated as basic factors and the second series, consume acceptance factors. They are as follows:

**Basic Factors**

1. Price to packager
2. Sanitary qualities (protection of product from contamination)
3. Resistance to impact injury
4. Efficacy of interior surface
5. Absence of handling problems

6. Space and other storage requirements in filling plan and in distribution including weight of package
7. Special features related to performance for packager

**Consumer Acceptance Factors**

1. Size
2. Ease of opening
3. Reseal features
4. Pouring qualities
5. Space saving in consumer's premises
6. Light protection
7. Transparency
8. Tamperproofness
9. Protection of contents from physical or chemical change
10. Physical characteristics of outside surface, including appearance
11. Ease of disposal
12. Special features related to performance for consumer

Some of these factors are not used in appraising certain types of containers. The reseal factor, for example, is ignored in appraising containers which are never to be stored after opening, containing a portion of the original contents. For some types of container, for example, a paper milk container, all of the listed factors may be needed in an appraisal.

**Rating the Container**

A procedure for rating the container on a numerical basis, in respect to each factor, must be evolved. The first step is to weigh the factors numerically in respect to each other; the second step is to evolve a procedure, including a mathematical formula, for computing the rating of a container in respect to each factor. These steps are to be taken with extreme care and deliberation by a panel or committee composed of specialists in technology, engineering, manufacturing, selling, and purchasing.

**Weighted Values of Factors**

After a complete discussion, the weighted value for each factor should be established by a majority vote of the committee. If the elements involved are thoroughly discussed, so that differences of opinion are effectually clarified, these decisions will usually be unanimous.

Evaluating the functional quality factors in this manner prevents the placing of unjustified emphasis on a particular factor which, when viewed in an isolated position, may appear to be endowed with great importance but which, when weighed carefully in relation to other factors, is of quite ordinary stature. Thus, the entire problem is viewed within its proper perspective.

In weighting the factors, one must not be misled by the fact that a very low level of competency of a container in respect to any one of the denoted qualities may be, but is not necessarily, a sufficient fault to make the container unsuitable for the purpose intended. The existence of this fact does not justify questions of (1) the degree of facility with which the critical minimum level of competency can be avoided and (2) the relative values of the degrees of improved quality which exist potentially within the realms of the different quality factors in respect to the cost of attaining the improved quality. For instance, if a quality improvement can be obtained in the realm of factor No. 2 for one third of the cost of obtaining a quality improvement in the realm of factor No. 5 and the two improvements are judged to be of equal value to the prospects of success of the container, on the basis of this consideration, factor No. 2 should have a weighted value three times as great as that of factor No. 5. A very high level for the critical minimum, such as will exist for factor No. 2 (sanitary qualities), may merely be regarded in the light of placing a higher value on a quality improvement in the realm of that factor.

The significance of a given factor may be different when applied to one type of container than when applied to another type. For paper milk containers, several of the factors have special significance, as indicated below:

*Basic factor*

2. *(a)* protection of the pouring lip
   *(b)* interior sterility
3. leak proofness
4. *(a)* taste or flavour contamination
   *(b)* flaking or peeling of coating
5. *(a)* extra machinery required (over and above that required for the type of container in present use)
   *(b)* utilization of present filling equipment
   (c) filing speed
6. transportation to dairy may or may not be included

*Consumer factor*

4. includes facility for separating cream from milk
9. includes cream absorption and adsorption

For successful operation of this system of grading, an explicit pattern must be designed by the committee for each factor and set forth in writing so that there is complete understanding by the committee members of all principles and tenets laid down in the pattern. In these expositions, not only the effects of various features upon the grades must be given but, also, the reasons why these effects are attributed to the particular features. Hypothetical patterns of grading principles for two basic factors and one consumer acceptance factor, applying to paper milk containers, will illustrate the type of grading pattern which seems best to serve the purpose.

***Basic factor no. 1—Price***

1. Price includes only cost (material, labour, overhead) plus 7½ per cent profit before taxes-no transportation or jobbing cost.
2. Price of $ 10.00 per M containers is given an ideal rating of 30 points.
3. Rating is calculated by equation (1)

$$N = 30 - 3(x - 10) \quad \dots (1)$$

in which N is the factor rating of the container and x is the price per M containers.

***Basic factor no. 4—Interior coating***

1. Ideal equals 10, Ten points to be divided evenly between (a) and (b) all paraffin coated containers rated 5 under subdivision (a).
2. Three-ply type end is considered least likely to produce flaking and is rated 4 under subdivision (b).
3. Type C end decreases rating from that of 3-ply end by 'h point because of increased liklihood of formation of fillets.
4. Folding of types D and E tops after paraffining and less rigidity of these tops counterbalance the better paraffin drainage of these containers and the more numerous places for fillets in the type C end. The result is equal rating for types D and E. (Because of less rigidity during service, types D and E tops decrease the rating from that of the 3-ply top by 2 points and from that of the type C top by 1½ points.)
5. Type F top has same rating as type D top.
6. Type G end decreases the rating from that of type C end by ½ point because of lower rigidity of type G end.
7. Type F bottom has same rating as type G bottom.
8. Use of type H top, with bottom filling, decreases rating by point from that of type C top and bottom because of increased danger of sluffing of paraffin in opening the bottom for filling.

9. Low coefficient of rigidity, increased area of folded top, and sewing of top causes reduction in rating by 2 points of type I container compared to type D.
10. Type J top decreases rating by 3 points from that of 3- top because of reduced rigidity and large number of places for fillet formation in type J top.
11. Cream baffle in container 8'/a" high decreases rating by 'h point and in container 6" high by 1 point because of increase in opportunity for formation of fillets.
12. Body 6" high decreases rating by 1 point from that of body 8" high because of greater opportunity for formation of fillets around top and bottom.
13. Cylindrical body is considered equal to rectangular body with ends of type C; ½ point better than rectangular body with ends of type F or G.
14. Type K top has a rating 3 points lower than the 3-ply top because of the forming operation on the neck in closing.

***Consumer acceptance factor no. 4—Pouring qualities***

1. Ideal equals 3.
2. All containers except hype I, type L, and short cylinder are rated 2.
3. Type L with round opening, short cylinder with opening near periphery of end, and type I are rated 1½; *type* L with extended opening is rated 2; short cylinder with opening in centre of end is rated ½.

Note on above patterns: Types C to L, inclusive, designate particular designs, all of which are familiar to the committee members. These types figure in the grading patterns of other factors in a similar manner to that of their use in the patterns illustrated above. All combinations of features are thus graded under all factors. The ratings for each combination, representing a specific container design, under the factors of both series, are added to produce a composite rating for the container. The container receiving the highest rating is judged to be the best for the purpose.

## Discussion of Grading Factors, Applied to Different Types of Containers

***Basic factor no. 1—Price to packager***

This is the one grading factor in the consideration of which the purchasing department of the container manufacturing firm plays an active part. This department must supply information on the availability and the price of materials. It may also assist the technical department in determining whether or not a material is a suitable one for the container by obtaining certain items of technical information from the supplier.

Information collected by the purchasing department is combined with facts developed by the technical department and the manufacturing department on the manufacturing operations necessary; the accounting department then determines the cost of material, labour, and overhead, then the evaluating committee adds such percentage as has been decided upon as a proper mark-up to obtain the price, excluding storage and transportation on the finished container and jobbing cost. The committee then applies its price factor formula and obtains the price factor rating for the container.

It may be well to enumerate here a few principles on the relative costs of packages in different functional classifications. Generally, a container of type A in the following list costs more than a container of type B of similar size in the same line.

| *Type A* | *Type B* |
|---|---|
| hermetic | nonhermetic |
| rigid | nonrigid |
| vacuum holding for moist products | pressure holding for dry products |

***Basic factor no. 2—Sanitary qualities***

This factor is concerned with the protection afforded by the container to the contents against micro organisms and against filth. Any container which, in storage, affords protection from microorganisms protects also from filth; the reverse of this statement, however, is not necessarily true. It is an easier and a simpler matter to protect from filth than to protect from microorganisms. The latter type of protection is required for a product containing more than about 20 per cent of water, which is not protected by sufficient salt, sugar, acid, or alkali to prevent bacterial growth, and which must be held in the container longer than the normal spoilage time for the product at its storage temperature. Sterilized canned foods, of course, are the outstanding example of foods in this category. Hermetic containers are necessary to afford this protection.

Foods sold at retail that do not have to be hermetically packaged are predominantly packaged in tightly closed paper bags, folding boxes, film pouches, pouches of various laminates, molded plastic containers, or nonhermetic metal or glass containers. A container of any of these types, as long as it is unbroken and unopened, effectively protects its contents from filth. Among foods in this category are some which are subject to bacterial spoilage but of which the distribution is so controlled that spoilage does not normally take place. Such products are salad dressings, meat, fresh and cured and fresh dairy products, including fluid milk.

*Fresh and frozen cured meat*

The packaging of fresh raw meat and of ready-to-eat cured meat during the last decade has been the outgrowth of self-service marketing because, in this style of marketing, packaging is essential to protect the meat from filth. In late 1953, 50.3 per cent of the members of the National Association of Retail Grocers were using meat selfservice marketing. Throughout the United States at that time, the meat departments of 9,500 stores, including 3,750 independents, were operating on a 100 per cent self-service basis. Transparent films are used for this packaging because consumers insist on seeing the meat they buy.

In respect to quality deterioration of self-service meats, the dealer has only one serious problem-appearance. This is primarily a colour problem. Dehydration is indirectly involved because this aggravates the colour problem. Selection of packaging materials is done almost exclusively on the basis of the effects of these materials upon the preservation of colour in the meat. For the first two days after packaging, a permeable type of package for raw fresh meat seems to be advantageous because, with oxygen in contact with the meat, the redpigmented oxymyoglobin is preserved on the surface of the meat. For this reason, films which are permeable to gas, such as cellophane (when wet) or cellulose acetate, are usually preferred. The former is generally. preferred because cellulose acetate, has a high water vapour transformation rate and thus permits rapid loss of moisture from the meat. For preserving red colour longer than 48 hours, a film of very low gas permeability, such as a "Pliofilm" or Saran laminate with cellophane, serve better than cellophane alone, especially if the package is sealed under vacuum. In these films, however, the colour of the meat during the first 48 hours after packaging is not as red as that in cellophane. Cured meats differ in this respect from raw fresh meats; the red colour of the former is best retained in the complete absence of free oxygen. For this reason, these items are sometimes sealed under vacuum in packages of cellophane—"Pliofilm" or cellophane-polyethelene laminate. Polythene alone, being quite permeable to gas and not being grease-proof, is not a satisfactory packaging material for meat. The colour changes just discussed for both fresh and cured meats are explainable on the basis of some unique chemical properties of the heme pigments.

Film breakage also presents a problem for the self-service dealer in meats; however, as a rule, the film outlasts the red colour of the meat. Simple films, like cellophane, usually are not required to last longer than two days. The laminates resist breakage better than cellophane alone. A help in preventing film breakage is found in the use of molded wood pulp trays. The tray absorbs moisture and maintains the shape of the package.

Some frozen meats are packaged in transparent films but this practice has not met with wide acceptance because the appearance of the meat is impaired by the formation of frost on the inside surface of the film wherever the film is not in contact with the meat.

*Fresh vegetables and fruits*

To a minor degree for sanitary protection of the product, but mainly for convenience, fresh vegetables and fruits are packaged, to a rapidly growing extent, in plastic bags. Polyethylene is very popular for this purpose because of its physical strength and transparency, combined with the fact that it is cheaper than cloth. About 40 per cent of polyethylene bags in 1954 are used for produce. "Pliofilm" (rubber hydrochloride) also is used for this purpose-much of it in "stretch wrapping." It is claimed that 65 pounds of "Pliofilm" will individually wrap a carload of lettuee. Produce packaged in these film materials is cooled after crating by vacuum treatment, promoting evaporation of free moisture from the surfaces of the produce.

*Properties of films*

The importance of various physical and chemical properties of film materials should now be apparent to the reader. To explain briefly how objective measurements of those properties are made, which are of greatest importance in the field of food packaging, we shall quote from a nontechnical exposition on the subject which was issued in January, 1954 by Chester Packaging Products Corp. as an "Informational Bulletin" on polyethylene. Editor, Gene Liberty. One section, that on "pH", is omitted.

**Age of Productive Packaging**

"The vocabulary of the testing laboratory has become part of the everyday speech of the converter. In previous years, price, supply, delivery, demand, and the conversion of existing equipment to the handling of new films were the major uncertainties that converters faced. Today, in our industry's Age of Protective Packaging, there are additional concerns. Flexible films are so longer thought of in terms of simple containers or wrap-around covers. The demands of our complex technology have caught up to them.

"We ask new questions: What is the burst, tear, WVT, tensile, grease-proofness? Are the pH, aging, and oxygen, transmission satisfactory? Will it stand up to sunlight and have a long shelf life? Both industry and government now draw up exacting specifications of physical and chemical properties that are required for packaging films. As competition intensifies, these properties often decide whether a sale is made or lost.

"This issue of the Bulletin is planned as an abridged and simplified guide sheet, to give readers without too much technical experience an insight into some of the methods of determining protective properties of flexible films. Although we verified outside sources and literature, most of our information is derived from the day-to-day practices of our own laboratory.

***Are published figures of physical properties reliable?***

"Generally, published figures of physical properties are the findings of conscientious laboratory testing, but standards, personnel, equipment, and methods vary from laboratory to laboratory often producing results that conflict. Small changes in a single test procedure can yield answers that are surprisingly far apart, and only under carefully controlled conditions can two laboratories duplicate their efforts.

Until the time that technical societies succeed in universally standardizing test procedures for flexible packaging, it is prudent never to accept published figures at value. Figures alone are misleading; they have significance only when all the details of the test procedure are known.

**Water Vapour Transmission**

"Water vapour transmission (WVT), sometimes called moisture vapour transmission, (MVT), is a measurement of the weight of water vapour that will pass through a given area of packaging material every 24 hours. WVT refers to the passage of water *vapour* and should not be confused with waterproof, which describes the resistance of a sheet to *liquid* water.

"In the test, an aluminium cup is filled with a desiccant, i.e., a substance that absorbs water readily. The packaging material to be tested is then sealed air tight to the top of the cup with wax. Moisture can only get to the desiccant by passing through the sheet, and the amount of moisture which passes through and is held inside will determine the WVT of the sheet. The entire cup assembly, containing the cup, desicant, packaging material, and sealing wax, is next placed in a humidity cabinet that circulates hot moist air. One of the widely used cabinets in this work is the General Foods Humidity Cabinet, operating at 100° F and 90 per cent relative humidity.

"Some of the moisture in the cabinet will penetrate the sheet and be absorbed by the desiccant. The cup assembly is weighed every 24 hours to determine the amount of moisture gained. When this gain becomes constant and the cup assembly becomes heavier by identical amounts for each 24-hour period, the test is concluded. The WVT is calculated from the daily weight gains and the known area of the sheet.

"A typical WVT value for 1% mil polyethylene sheet tubing might read 0.94 gm/100 $in^2$/24 hrs. This means that at 100° F and 90 per cent relative humidity, 0.94 gms of water vapour will pass through a polyethylene sheet 100 square inches in area every 24 hours. Some laboratories report WVT in gm/$m^2$/24 hrs., where $m^2$ is square meters.

**Bursting Strength**

"A sample is clamped between the two ring shaped jaws of the testing machine and rests above a rubber diaphragm contained in the lower jaw. The diaphragm is connected to a hydraulic system and can be blown up like a piece of bubble gum by increasing pressure on the system. As more pressure is applied, the specimen, constantly being pushed by the expanding bubble, will burst. A dial gauge records the exact pressure which bursts the sample.

"Most laboratories have a Mullen Burst Tester and "What's the Mullen?" is frequently used to request information about bursting strength.

"Bursting strength is reported in either points or pounds per square inch; both units are approximately equal.

**Tensile Strength and Elongation**

"Although several machines made by different firms are used to determine tensile strength and elongation of thin films, a general test procedure can be outlined to describe all of them: Samples are cut in long narrow strips, for example, 7" × 1", and clamped between two jaws. A constantly increasing load is applied to one jaw moving it away from the other. The sample, which is held between the jaws, is literally pulled apart. First, it starts to stretch, and when it can stretch no further, it breaks in two. The minimum load required to break the specimen determines the tensile strength, and the amount of stretch just at breakage determines the elongation.

Tensile strength is reported in pounds per inch width (where width refers to the width of the simple) or in pounds per square inch. When pounds per inch width is used, the term breaking load is often substituted for tensible strength. Elongation is reported in inches or percentage. To illustrate, if the original distance between the clamps is 5" and the distance at breakage is 20", the elongation equals 15" or 300 per cent.

## Internal Tearing Resistance

"The simplest way to determine internal tearing resistance is on an Elmendorf machine. The sample is clamped between two sets of jaws, one movable and the other fixed. A pendulum is attached to the moving jaw, and when it is released, it plunges downward, carrying the jaw and tearing the sample. A pointer and dial on the pendulum indicate the internal tearing resistance.

"The operation of an Elmendorf may be compared to clamping a sample in a bench vise and tearing the projecting piece with your fingers. The vise is the fixed set of jaws, your fingers are the movable set, and your forearm, which supplies the motion, is the pendulum.

"For some films, the Elmendorf is unsatisfactory, and more complex technics of determining the tearing resistance are necessary.

"Internal tearing strength is expressed in grams per sheet or pounds per inch.

## Grease Resistance

"Greaseproof tests attempt an accelerated comparison of the rate at which oils and greases, such as those found in foods and industrial preservatives, penetrate packaging material A 4" × 4" sample is placed on white paper, and a measured volume of test reagent is poured on top of the sample. The regent is generally colored turpentine or a solution of grease in turpentine. Periodically, the white paper is examined for small stains to see if the reagent has penetrated the sample. As soon as a stain is discovered, the test is terminated.

"The time elapsed from the start of the test to the appearance of the first stain is known as the transudation time. Depending on the reagent and the severity of the test, acceptable transudation times vary from 15 minutes to 24 hours.

**Gas Permeability**

"The ability of a packaging film to permit or prevent the transmission of a gas frequently decides whether the product enjoys a long or short shelf life. Typical problems related to gas permeability are oxygen induced deterioration of cheeses, dried whole milk, and fats; product contamination through absorption of foreign odours; excessive loss of desirable odours from substances like bath salts or lotions; and tarnishing of silverware. At times, high permeabilities are desirable as in the "breathing" of fresh produce.

"In the test procedure the film is sealed across an opening in a vacuum chamber equipped with a pressure gauge. The gas, which may be air, oxygen, hydrogen, carbon dioxide, etc., is passed through the film into the chamber. As the chamber slowly fills with gas, the pressure rises and gauge readings are made at time intervals. When a leveling off period occurs and the same volume of gas enters every hour, the test is completed. The film will exhibit different permeabilities to different gases.

"Gas permeabilities are reported in cc/100 $in^2$/24 hrs. (cc is an abbreviation for cubic centimetre, the scientific unit of volume. One quart contains approximately' 943 cc.).

"During its life, a package must resist a combination of atmospheric influences which eventually age and destroy it. These influences are made up of varying proportions of humidity, heat, cold, sunlight, pressure, and gases. In an accelerated aging test, the film or package is subjected to harsh extremes of heat, cold, humidity, etc., in order to accelerate the changes that would take place under the less severe conditions of normal service.

"A typical test is to alternatively expose a sample to wet and dry heats such as 24 hours at 160° F. in a dry oven. This wet heat/dry heat cycle is repeated as often as the test conditions require. If necessary, other factors like low temperatures, ultraviolet light or oxygen exposure are introduced into the cycle. After each cycle is completed, the film or package is examined for product deterioration, changes in weight and dimensions, dulling, crazing, warping, and discoloration.

"It is difficult to predict the shelf life of a packaged product solely from the results of accelerated aging tests unless some correlation can be made based on years of field testing. Many laboratories agree that shelf life is best determined by periodic examination of packages that have been placed on shelves in different parts of the country or world. This is an unfortunately long but always reliable procedure. However, accelerated aging tests are valuable in comparison testing, where different films are simultaneously put through aging cycles to ascertain which of them will best withstand the corrosions of the atmosphere."

The excellent plasticity of polyethylene film at low temperatures makes the film well suited for packaging frozen foods when the film does not have rigid support against breakage. An interesting new use of polyethylene, laminated with cellophane with a special low temperature adhesive, is in pouches for packaging individual servings of orange juice concentrate (2½ ounces per package). The pouches are heat sealed.

***Fluid milk***

Probably the most widely used food among those which are distributed in a nonsterile state and which are packaged for sanitary protection, is fluid milk. Packages used are glass bottles with paraffin coated paperboard or foil closures and paperboard containers coated with paraffin or plastic. To measure the sanitary protection that is afforded to the milk the package is evaluated from the standpoint of two features, namely, (1) protection of the pouring lip and (2) interior sterility. Although fluid milk is a nonsterile product, the number of viable bacteria in it is carefully controlled. Maximum numbers are specified in control regulations which are enforced by public health officials. Therefore, it is required that the interior of fluid milk containers, before filling, be either sterile or nearly sterile so that the bacteria count of the milk will not be noticeably increased by bacteria from the container. A container is regarded as sterile if it can be rinsed with sterile water without imparting any viable bacteria to the water. Sterility of glass milk bottles is ensured by thoroughly washing the bottles just before filling with a hot antiseptic solution and rinsing with sterile water. With the most improved technic in washing, a detergent is added to the washing solution automatically, keeping the solution at constant strength. With paper milk containers, the final treatment with hot paraffin or hot plastic produces the sterile condition although, prior to that, the paperboard, of which the containers are made, is produced from virgin pulp by the sulphite process and under regulated calendering conditions which ensure a very low bacteria count in the board and complete absence of bacteria of types that have sanitary significance in milk. The paraffin or plastic coating of some containers is applied in the dairy plant in the container fabricating operation just before the containers are filled; other containers are coated in a central fibricating plant before being shipped to the dairy. In the latter case, each container is tightly closed immediately after the coating is completed to prevent the entrance into the container of bacteria or other contaminants during shipping and handling prior to filling.

By "protection of the pouring lip" is meant the coverage immediately after the container is sterilized and immediately after it is filled, of all surfaces near the pouring opening, which might be touched by the milk as it is being poured. This is to prevent these surfaces from acquiring unsanitary contaminants which would be imparted to the milk when it is poured. It is accomplished on glass bottles by having the lip of the bottle overlapped either by a specially designed combination plug and cap or by a separate hood. On paper containers, the coverage is afforded by either an extension of or an attachment to the plug, which is attached to the container in a hinged manner.

To avoid holding milk too long, that which is left on the shelf after 24 hours is removed and applied to other use than that of fresh milk.

### *Salad dressing*

A stock rotation programme of a different sort than that used with milk is employed with salad dressings and products of a similar degree of stability. Principally, glass containers are used for such products and code marks are put on the labels, either by printing or by perforating, to indicate the date of manufacture of the product. The perforating method has generally been used in the past because of the difficulty of printing on varnished labels. By a new technic, printing is now being done, providing both speedier operation and improved legibility over the perforated code. The shelf life of these products, of course, is much longer than that of fresh milk, being ordinarily counted in terms of weeks.

### *Basic factor no. 3—Resistance to impact injury*

It would be a happy circumstance if every container could be made physically strong enough to resist breakage under any impact. Flexible packages, folding boxes, glass containers, in fact, practically all containers except those made of metal, as a rule, show little visible effect of impacts which are not severe enough to break the containers; metal containers, however, may receive dents, which are permanent disfigurations, from impacts which fail both to break open the package or to render its seal ineffective.

Evaluation of the probable severity of impacts which a container will have to suffer is a necessary step in selecting the type of container to use for a product. Not only must be material of which the container is to be made, along with the design and the type of construction, be considered, but also the nature of the product to be contained. With the liquid product, the problem of the effect of impact upon the container is more critical than with a dry product because liquid spilled from a container usually destroys the usefulness of other containers with which the liquid comes into contact. When a dry product is spilled, surrounding containers often are not seriously affected. Pouches of flexible material are often not affected by impact until the impact squeezes the pouch to the bursting point. In paperboard containers, rectilinear containers are more resistant to most types of impact than curvilinear containers, although in some cases of point impact, the curvilinear might show superior resistance. Multiply folded ends lend strength to paperboard containers but such ends are not as strong as a single ply end of equal thickness unless the piles of the folded end are fastened together with adhesive.

In metal cans, which are the most resistant of the retail food containers to physical abuse, the hermetic type of container presents the most serious problem. When a hard object strikes a double seam a blow, such as may occur when a filled can is rolled down a steep decline and suddenly stopped by coming into contact with a metal barrier, the conformation of the seam and its gasket may be sufficiently disturbed to cause a leak which would admit air to the can and might even admit spoilage bacteria. Such treatment is not likely to break a well soldered sideseam; however, a current move to reduce the use in cans of metals which become critical in times of emergency, poses a question as to the resistance offered by solderless sideseams to the injurious effect of impact. For certain nonfood products, cans with "doped" or cemented sideseams have been used for a number of years. Such cans, if the sideseam compound is innocuous when in contact with foods, are satisfactory also for foods that are not processed for sterilization. A sideseam compound that is suitable for use with foods, and which will withstand heat processes for sterilization, however, has not been brought to a practical stage.

Compounding a sideseam is a much more difficult problem than compounding a double seam for the reasons that:

1. The sideseam, being straight and being made with one less fold than the double seam in the stock forming the outer hook, does

not have the assistance of the curvilinear shape and of the extra fold of stock which the double seam has in preventing unhooking and in preserving the orientation of the hooks in respect to one another,

2. The sideseam cannot readily be rolled to tighten it after it is formed.
3. The sideseam will not stand tightening to the same degree as the double seam on account of the greater tendency of the former to become unhooked from tightening, due to the mechanical characteristics mentioned in reason No. 1,
4. The sideseam cannot be hooked over its entire length; at each end, it is simply lapped to allow for doubleseaming,
5. The sideseam is more susceptible than the double seam to severe deformation by impact, due to the mechanical characteristics mentioned in reason No. 1. For these reasons, it is very difficult to formulate a workable compound or cement containing a sufficiently large proportion of inert insoluble substances to hermetically seal a sideseam and to possess reasonable resistance to leakage as a result of deformation of the scam.

While the use of compound in sideseams saves solder, more important is the saving of tin on the plate that is made possible.

For food products of a noncorrosive nature, the only reason that a tin coating is required on the plate is to make it possible to solder the sideseams of the cans. Welding of sideseams on a commercial scale is possible but this is more expensive than compounding. Reasons for this will be stated later.

***Basic factor no. 4—Efficiency of inferior surface***

The interior surface of any food package must be compatible with the contents, that is, it should be of such a nature that any chemical or physical reaction that takes place will have a minimum effect upon the quality of the food. Most materials of which food containers are made are relatively inert chemically. With food products in the dry state, the chemical nature of the container material is of little moment because the food itself is inert. Paper and synthetic plastic are the most commonly used materials for these foods-untreated when the container is to serve for a short time only or when change in moisture content of the food during storage is immaterial. Where it is necessary to protect the food against change in moisture content, plastic material, if not naturally impervious to moisture, or paper, is treated with wax or other substance, often plastic, which renders the package moisture-resistant. While this has nothing to do with chemical reaction between the food and its container, it may be said to be concerned with physical reaction between them, in which the container material either takes from the food or imparts to it water.

Foods in the range of moisture content of 20 per cent and higher are in a class which react, both physically and chemically, with the container. It is with these foods that the major problems dealing with the inside surface of the container exist. Of course, the necessity of protecting the food, either of dry or moist type, from absorption of foreign odour or flavour or of toxic fumes is always present. The problem involved here is concerned with the inside surface but is concerned even more with the packaging material in its entirety. The body of the material must contain nothing which is capable of imparting to the food either toxicity or foreign odour or taste; no food packer and no container manufacturer would think of attempting to protect a food product from dangerous or undesirable elements existing within a packaging material by superimposing an innocuous surface coating on the inside of the package to serve as a mask for the dangerous or undesirable elements. Such coatings are employed, however, to shield food from a container material which is capable of reacting chemically with the food in a manner which has only economic significance, accompanied by no health hazard and by little or no quality deterioration of the food.

Under Principles of Grading, is given a pattern of grading principles which illustrates how the inside surface of one type of container, the paper milk container, might be evaluated. This pattern is simplified by acceptance of the assumption that chemical reaction is no problem with either the paperboard or the coating material; thus the rating is concerned only with physical reaction between the container and the milk, a liquid product. Basically, this constitutes a rating of the efficiencies of different types of coatings in imparting moisture resistance to containers of various designs when the coatings are applied in various ways and in various stages of fabrication and handling. If container materials are being considered which are not known to be free from toxic ingredients or to be chemically inert in respect to milk, the problem of evaluation would have numerous additional ramifications.

*Problems with steel*

The so-called "tin can" is a container made of steel plate on which is a very thin coating of metallic tin. The necessity to conserve tin during World War II led to the practice of using thinner and thinner coatings of tin on plate used for food containers. The technology involved in this movement is very complex from the standpoint both of can manufacture and can use. Progress has been rapid, however; cans can be made without tin that are satisfactory for nonsterilized foods as well as for some heat processed foods. The cans must be coated on the inside with an organic material which is inert in respect to the food product to be held and on the outside with a substance which will prevent rust. When the sideseams are welded, these coatings, at least on the areas adjacent to the sideseams, must be applied after the body of the can is formed because, when applied previously, they do not withstand the heat of welding. This operation, along with the fact that maximum speed in welding is substantially less than customary soldering speed, makes the cost of welded cans greater than that of soldered cans, notwithstanding the fact that no tin is used in the welded cans. Thus, while welded cans are used commercially in Europe, they have not yet been found to be economically feasible in the United States. The industry constantly seeks to remedy this situation.

Compounded, or cemented, sideseams can be made at high speed and with no accompanying problem in relation to coatings. The formulation of a sideseam compound for cans of food to be heat processed, however, is not perfected. Sideseams of tinless cans, made of CMQ (can making quality) of black plate, can be soldered at high speed but often with injury to enamel coatings and with corrosion of the plate due to use of flux having highly corrosive properties. Moreover, adhesion of enamel coatings to this plate is very difficult to secure. Black plate can be made more receptive to enamel coatings by a chemical treatment but all such chemical treatments now known increase the difficulty of soldering. Thus, soldered tinless cans also are not yet feasible for wet food products.

Steel plate can be used successfully in containers for wet food products only when the steel is covered with a coating either of a metal of low chemical reactivity, such as tin or aluminum, or of an organic substance, or of both together. The organic coating substances may be composed of either synthetic or nonsynthetic materials. They may be mixtures of thermoplastic resins and plasticizers which are applied in the melted state and allowed to solidify in cooling, or they may be substances that are baked to "set" them after they have been applied as coatings. Coating materials that require baking are formulated either as cooked mixtures of oil and resin or as uncooked solutions of resin and plasticizer. The coatings problem exists with the all-metal can and with metal caps of glass containers. There is no surface problem with the glass portions of food containers.

Chemically, tin used for food cans must contain no more than the amounts of impurities specified by Federal Agencies as maximum tolerances. Thus, toxically, tin provides a safe inside surface for food cans. Were it possible to produce perfect coverage of steel with tin, there would be practically no chemical reaction between food products and metal. Because of exposure of minute areas of. steel, however, galvanic couples of tin and iron are established in contact with the food product, which serves as an electrolyte, and a solution of either tin or iron results. The rate at which this solution take place depends upon the oxidation-reduction potential of the electrolyte, the temperature, and the presence or absence of other metallic elements in the tin plate. Low concentrations of certain elements, some metallic, some metalloidal (silicon, phosphorus, chromium, nickel, copper, molybdenum are of greatest significance), cause the reaction to go slowly. Only within the last 25 years, during which the cold reduction process for tin mill sheets has been in use, was it learned how to produce the proper degree of hardness and proper surface characteristics for satisfactory fabrication of cans without the use of comparatively high concentrations of metalloids in the steel. Thus, cold reduction and temper rolling of steel sheets have led to the solution of many problems associated with corrosion of tin plate by foods.

There are two ways of controlling the kind of corrosion that results in reduced shelf life for canned foods; sometimes one is used, sometimes the other, depending upon prevailing conditions, both economic and political, as well as upon the nature of the product being packed. The first method is to use a heavy enough tin coating to retard corrosion This is always a feasible method provided the necessary tin is available, for, notwithstanding the presence of minute exposed areas of steel through pores in the tin coating, tin provides protection against corrosion by retarding the solution of the iron, and the extent of retardation of solution in proportional to the thickness of the tin coating, within the range of thickness that is used on plate for cans, namely, up to about 2 pounds of tin per base box (62,720 sq in. of coating) of plate. While this procedure is always feasible from the standpoint of preventing destructive solution of iron, it is not feasible with some products from the standpoint of the preservation of a desirable appearance of the product or of the interior surface of the can.

Some products cause the surface of tin plate, regardless of the thickness of the tin coating, to take on an undesirable appearance, due to chemical reactions between the product and the metal. With low-acid products, which liberate considerable quantities of sulphur-bearing compounds when they are heatprocessed, dark-coloured sulphides are formed. Prominent in this class of products are corn and seafoods, particularly shellfish. The more acid foods produce a mottled surface appearance on the tin plate, sometimes referred to as spangling or etching. Foods containing red pigment lose much of their colour when held in contact with plain tin plate, that is, plate on which the tin is not covered with a coating of lacquer or other organic material. Because of these effects of holding in plain tin cans, the inside surfaces of cans for these products are lacquer coated, regardless of how heavy the tin coating on the plate may be.

Baked lacquer (or "enamel," as it is usually called) coatings on the interior surfaces of food containers retard corrosion of the tin plate; thus, the second way to increase the shelf life of a product is to use lacquer coatings. This is true especially with tin plate of higher temper which is used for the loss corrosive low-acid, products. Thus, for products that require enameled cans to ensure a satisfactory appearance of either the food or the inside surface of the container, the enamel coating may not only perform that purpose but also extend the life of the canned product. Lacquer coatings are often used however for the sole purpose of obtaining satisfactory shelf life when the tin coating is too thin to perform this purpose alone. With the relatively nonreactive products, such as corn, peas, meat, and fish, therefore, as well as with the moderately reactive products, such as peaches, pears, citrus fruits, pineapple, tomatoes, green beans, and beets, lacquer is substituted for tin when conditions make it desirable or necessary to conserve tin. (In beets, it preserves colour, also.)

Conservation of tin on a major scale in the manufacture of tin plate has been made possible by the building, within a period of just 15 years, of capacity to produce electrolytically coated tin plate at a rate equal to 90 per cent of the world's maximum rate of production of tin plate. By the older method of coating, known as the hot-dip method, it has never been found practicable to produce plate having less than 1.25 pounds of tin per base box, whereas, by electroplating, the minimum limit of tin for satisfactory coating is nil. Electroplating equipment now in operation in the countries of the Western World consists of 41 production lines, of which 33 are in the United States,

3 in Great Britain, 2 in Canada, and 1 each in France, Belgium, and Brazil. The United States production capacity of electrolytic plate, based on No. 50 plate which carries .5 pound of tin per base box, is about 3,845,000 long tons and the capacity of other countries is about 910,000 long tons per years.

*Properties of enamel coatings*

Studies are being made continuously of the properties of solvents, oils, resins, and plasticizers used in enamel coatings to ensure that the coatings are chemically inert to the food products with which they are to be used and are incapable of imparting any harmful or disagreeable characteristic to the food. Among the properties of a coating which are most important to ensure that it will function as intended is one which depends upon both physical and chemical factors. It is the property of adhesion. The ability of the coating to adhere to the plate depends not only upon the composition of the coating material and the procedure followed in applying the coating, but also upon the surface characteristics of the plate, which, in turn, are determined by the composition of the steel, the finish of the steel, and the manner in which the tin coating is applied and finished. These characteristics also demand constant study to determine the effect of every variation in plate fabrication procedure upon the performance of coatings. Failure of a coating to perform properly means loss of food, which has not only been produced, but also canned.

There are three ways of applying tin electrolytically, differing basically from each other in the type of electrolyte used. The essential steps of the operation, including cleaning, plating, fusing, oiling, etc., are to a large extent, performed differently in the various methods. Any one of these variations may importantly affect enamel adhesion and, finally, the baking of the lacquer itself is a most critical step. The baking technic must be correlated with all of the other factors to which reference has been made including, of course, the end use of the container.

Another type of plate which is coming into use on an important scale is differentially coated, that is, plate with a thicker coating of tin on the one side than on the other. The purpose of this plate is to provide a thick coating of tin on the inside of the can for protection against corrosion by the food and a thin coating, probably a No. 25 coating, on the outside, which is ample for protection against rusting under normal storage conditions. It is expected that, generally, in differentially coated plate, the lighter coating will be 25 pound per base box or less; the heavier coating generally 1.0 pound per base box. Differentially coated plate is used to make cans for processed foods; up to this time, most of these have been for tomato juice, although as more differentially coated plate is produced, it will be used for other products of moderate or high corrosive activity.

The problems with steel, including those on organic coatings, have been discussed in direct reference to cans. The discussion applies also to metal closures for glass containers, except that part which pertains to the fabrication of seams. Problems of corrosion in metal caps differ but little from those in cans, although measures which have been found to be effective in retarding corrosion of the steel in cans seem not to be so effective for closures. Kohman believes that the mechanism of corrosion in metal caps includes the passage of hydrogen ions, and possibly of sodium and potassium ions, through the seal of the jar, enabling a galvanic couple to be established with its cathode outside the jar and its anode inside the jar. The cathode, being the raw steel cutedge on the periphery of the cap, is of comparatively large area; therefore, corrosion is accelerated.

***Basic factor no. 5—Absence of handling problems***

Most modern food packaging operations employ machinery for carrying out the various steps. The relative quantity of machinery required, and the relative simplicity or complexity of the machinery required by one container compared to that required by other containers, measure the advantages and disadvantages of that container in respect to others that might serve the same purpose. This characteristics, of course, pertains primarily to filling, sealing, and conveying machines and accessory equipment.

The conditions surrounding the projected use of a container often are important in this phase of a container evaluation. If the container would displace another container in the application for which it is considered, for example, in the replacing of one for the other among metal, glass, and paper containers, the question would be faced as to how much new handling machinery would have to replace the old but, if an entirely new operation were under consideration, the question would be whether or not a new installation for one container would cost less than, or have other advantages over, a new installation for another container. Sometimes the ease with which accessory devices or gadgets which may be considered important to the operation can be incorporated into the handling equipment is considered in weighing relative advantages of equipment. For instance, a magnetic attachment for removing tramp metal from the food being conveyed to the filler or an x-ray attachment for detecting foreign substances in a filled container or for checking the fill of the containers may have a substantial bearing on overall plans for machinery. Or, where rigid containers are involved, the requirements of machinery for uncasing and washing containers may have a substantial influence on the equipment layout.

Speed of filling and sealing naturally weighs heavily in consideration of the handling machinery factor in container evaluation. With one-line filling and sealing speeds approaching 800 per minute, competition amongst different types of containers is keen and correct decisions on these matters are important.

Machinery now is available even for filling 55-gallon drums with liquid automatically at a rate of two drums per minute.

***Basic factor no. 6—Space and other storage requirements in filling plant and in distribution, including weight***

The material of which the containers are made, the shape of the container, and the location of the container fabricating operations are the attributes which decide the container rating under basic factor No. 6. The greatest spacesaving step for the packing plant that uses rigid containers is that of receiving of container-making material in the form of container blanks and carrying on the final fabricating operation in the packing plant, preferably in tandem with the filling step. This system is used extensively with paperboard containers and to a minor extent with metal containers. Nonrigid containers, being collapsible, may be prefabricated without a sacrifice of storage space in the filling plant. Of the rigid containers that the prefabricated, those of rectilinear outline have an advantage over those of curvilinear design, such as cylindrical containers, from the standpoint of space requirements. Further, from the standpoint of both space requirement and weight, metal containers have an advantage over glass containers and paper containers have an advantage over metal and glass containers.

***Basic factor no. 7—Special features related to performance for packager***

A container may have attributes which, while they are unessential for excellent performance on the part of the container, do add substantial esthetic or utilitarian value to the container. Such an attribute might be one that permits the application of a special type of label which cannot be used on other containers intended for the same purpose.

This label might have unusual attractiveness and it might also enhance the utility of the container in some way that is "beyond the call of duty". The degree of acceptability of printing or lithography by a film may also be a case in point; for example, polyethylene may present more of a problem in this respect than some other film material.

Metal foil may be laminated in an overwrap or be used as a tray for the purpose of protecting against transmission of gas and moisture, but foil used in these ways is said also to accelerate freezing and thawing of the food. This is an added point of merit.

A canner, using conventional methods of processing, may contemplate converting his operation at a future time to one of aseptic canning. In choosing a container for his product, he would consider attributes which make the container adaptable to the aseptic type of operation.

*Consumer acceptance factor no. 1—Size*

The attribute of size has a bearing on the rating of a container only when a type of container which is customarily in use for certain purposes is being considered for a new use. The container would currently be available in certain sizes and it would be necessary either to use the container in one of the available sizes or to assume the cost of initiating its manufacture in a new size. The decision must be made after consideration of the consumers' desires.

*Consumer acceptance factor no. 2—Ease of opening*

In its bearing on consumer acceptance, easy opening carries great importance. Certainly, it receives as much attention in the development of a container as any other consumer acceptance aspect. Without undertaking a review of the methods of opening containers, let it be said that improvements in cans to facilitate opening occur quite infrequently; those in glass jar closures perhaps somewhat more frequently, and those in fibre and plastic containers are remarkably frequent. Screw caps and crown caps are used on both glass and metal containers; also screw caps are used on some fibre container. These are removed with comparative ease by gadgets that are common household articles. The most conveniently opened hermetic can is the keyopening scored can and the most conveniently opened hermetic glass container, aside from the rubber gasketed screw cap container, is the press-on pry-off ruber gasketed side and/or top seal container. Devices to facilitate the opening of fibre and plastic containers are many and of widely varied design.

*Consumer acceptance factor no. 3—Reseal features*

Milk containers must be provided with a means for resealing. On paper containers, a part which is hinged onto the container serves as a reclosure plug. The desirability of resealing many other products helps to determine the type of container that is used for these products. For coffee, the key-opening "collar type" can is generally used, the "collar" being a band encircling the inside periphery of the can, over which the side wall flange, or skirt, of the top end, after its removal, fits for reclosure. Where glass containers are suitable for hermetically packed products, the press-on cap provides a very satisfactory reclosure. Special reclosure can tops made of aluminium are now offered for sale to the consumer for use on cans after the regular top has been removed. Screw caps of either metal or plastic serve admirably for reclosing the containers fitted with them. These are practically never used commercially for containers that are to be hermetically sealed.

*Consumer acceptance factor no. 4—Pouring qualities*

A sharp edge serves best as a pouring edge because it cuts off a stream of liquid abruptly, preventing dribbling and running down the outside of the container after pouring has ceased. A sharp edge which is likely to be touched in handling, however, is hazardous because it may cut the hand. Metal and paper edges often can be turned in so as to leave a sharp edge where it can serve to cut off a stream in pouring and still not be exposed to contact with the hands.

In milk containers, if facility to separate cream from milk is regarded as desirable, the container will be evaluated in that respect under Pouring Qualities.

*Consumer acceptance factor no. 5—Space saving in consumers' premises*

Shape of the sealed container decides everything in this phase of container evaluation. Rectilinear outline, combined with flat ends, is the ideal and merits the highest rating. Other shapes merit lower ratings, which depend upon the extent of departure from the ideal.

*Consumer acceptance factor no. 6—Light protection*

Light accelerates deterioration of quality in some food products; for example oxidation reactions which impair the flavour of milk and the colour of cured meat. A container that protects such a product from light of the wave lengths which are injurious to it has special merit for that product. Amber glass milk and beer bottles and opaque paper milk containers fall in this category.

*Consumer acceptance factor no. 7—Transparency*

As implied in the last paragraph, transparency at times is undesirable from the standpoint of quality preservation of the food product. In fact, transparency is probably never helpful to quality preservation. Generally, however, it is not harmful and, in view of the fact that the consumer is always pleased when he can see what he is buying, the transparent package is considered to have a sales appeal which the nontransparent package does not have.

*Consumer acceptance factor no. 8—Tamperproofness*

All containers holding sterile food may be said to be automatically tamperproof because, if the seal is broken, spoilage sets in as a result of bacterial contamination. With nonsterile foods, however, a means of betraying tampering with the protective features of the container is desirable to avoid the possibility of using food into which filth may have been introduced. A container for such foods is less than adequate unless it is so constructed that, if the package, while containing the food, had been opened and reclosed, that fact will be revealed by the exterior appearance of the container.

*Consumer acceptance factor no. 9—Protection of contents from physical or chemical change*

The degree of protection from physical and chemical change afforded by a container to its contents is indeed a major point for consideration by one who is selecting a container for his product. However, the distinction in this respect between two types of containers, both of which are suitable in other respects for holding the product, is not great. It is usually a question of the length of time during which protection will be effective. Furthermore, the production costs of the *container* which is capable of giving longer protection usually are no more than the costs of the one which gives protection for a shorter period of time. The difference lies primarily in "know-how," that is, knowing how to make the better container.

Except in so far as light effects enter the picture, glass, being inert to foods, give excellent protection from physical and chemical changes. Problems involved in chemical reactions between foods and the metal caps of glass containers and between foods and metal cans were discussed under Basic Factor No. 4. A problem associated with glass containers, which was not discussed in that section, is that of air seepage through the seals of metal caps. Air *entering* the container during storage is responsible for producing *browning* or graying of the portions of food near the caps. Kohman attributes this *darkening* to ferric iron compounds which are produced by the oxygen that seeps through the seal. Livingston *et al.* found that the amount of discoloration at times could be correlated with the amount of iron present but found that a similar situation existed with other metals. They found also that gradations in the amount of discoloration in different portions of the contents of a glass container are not necessarily accompanied by corresponding gradations in the amount of iron present. Eolkin developed a method of measuring very small changes of gas pressure, such as those that occur in container when small quantities of air diffuse through the gasket. The extensive studies of van Amerongen and others on the permeability, diffusivity, and solubility of gases in rubber have pointed the way to the elimination of the discoloration trouble in glass containers through the substitution of rubbers of low gas permeability, such as butadience-acrylonitrile polymers and isoprene (methylbutadiene)-acrylonitrile copolymers, for GR-S type rubbers, which were used for gaskets until recently.

An effect on the colour of food, corresponding to that occurring in glass containers, is not apparent in cans-perhaps because of the relatively small cross sectional area of the gasket after it is compressed and tightly confined in the double seam. Other factors that might contribute to the prevention of the passage of air through a double seam are:

1. Relatively large areas of compressed metal-to-metal contact in the double seam, which probably retard the gaining of access of air to the gasket and
2. Increased energy of activation for diffusion in the rubber, which might result from high compression of the rubber in the double seam.

For frozen foods, since oxygen accelerates enzyme action, and for dehydrated foods, since both oxygen and moisture accelerate this action, there is a growing practice of using containers that protect the foods from these factors. Protection provided for these foods is now almost universally equal at least to that provided by a metal foil laminate and the use of hermetically sealed cans and jars is growing rapidly.

Interest also is growing in the practice of in-package desiccation of dehydrated foods, where the dehydration of the food is continued after the package is sealed, through absorption of moisture by a desiccant placed in the package. Moisture resistant package must also be used to prevent loss of moisture by foods of comparatively low moisture content, in which a certain minimum of moisture is necessary for the preservation of quality. Examples are unpopped pop-corn and dried fruits. For popcorn, sealed cans are used; for dried fruits, foil laminates and other materials of comparable protective ability are used.

The custom is growing of obtaining double protection by enclosing from two to six small fibre or plastic packages of food in a single wrapper made of the same material as that used in the smaller packages. Materials frequently used for such packages are polyethylene film and metal foil laminates.

*Consumer acceptance factor no. 10—Physical characteristics of outside surface, including appearance*

The outside surface of a retail food container needs to please two senses, sight and touch. To the touch, it should not give the impression of being greasy or sticky and, to the sight, it should be attractive. The attractive package on the supermarket shelf gets preferential attention of the family buyer. Label designs printed with fluorescent ink, which glows when illuminated by black light, are not uncommon and food manufacturing and marketing companies have designers constantly at work improving the colouring and the designs of their labels. Under Basic Factor No. 4, the outside surface of the container was mentioned in association with the container manufacturer and food packer's problems involving the inside surface. Protection of the outside surfaces of metal containers from rust is necessary, not only from the standpoint of preserving the food, but also from that of pleasing the consumer. A chemical treatment of untinned steel plate to prevent rust has become strikingly more effective, even within the past year or two, than it was previously and, where soldering is not necessary chemically treated plate can be used in containers with assurance that outside rust will not be problem even though the surface is not protected by organic coatings or lithography.

*Consumer acceptance factor no. 11—Ease of disposal*

In these days of regular garbage and trash collections in cities, both large and small, the disposal of empty food containers in the home presents little difficulty. It continues to be true, however, that the most easily disposed of container is the one that can be burried. These include wood, fibre, and most plastic containers. Tin-plate containers rust away in time; glass containers never disintegrate except by being ground to bits.

*Consumer acceptance factor no. 12—Special features related to performance for consumer*

Certain features of a container, intended for one purpose, may increase its usefulness to the consumer in some quite unrelated respect. For instance, a durable reclosure on a glass or metal container makes the container permanently useful in the home; or the rigidity or special shape of a container may enable it to withstand rough handling in shipment, or make it adaptable to dispensing in a vending machine, thereby increasing its convenience to the consumer.

**Final Step in Container Evaluation**

To illustrate the final step in the process of evaluating containers in the manner proposed in the preceding pages, a tabulation of grades under the different factors for six container variations taken from a group of about 60 which were include in a study that resulted in the selection of container No. 30. It is noted that, on consumer acceptance alone, container No. 55 has the highest rating and, in the category of basic factors, container No. 43 has the highest rating.

# 12

# Recombinant DNA Vaccines

Although the development of methods in genetic engineering has been rapid and their subsequent usage has, in part, revolutionized basic biomedical research, to date the number of marketable, industrial applications have been few. The field of veterinary biologics has been the primary benefactor of the limited number of genetically engineered products. Several recombinant DNA (r-DNA) vaccines used to prevent enterotoxigenic *Escherichia coli* disease in livestock animals are currently available world-wide and a vaccine for foot and mouth disease is available in limited markets.

## Rationale

The use of r-DNA technology in vaccine development provides several obvious and in some cases necessary advantages over previous methods. In general, bacterial vaccines have been divided into two groups: bacterins and toxoids. As their name implies, bacterins are prepared from whole bacterial cells. The cells may remain viable or treated chemically or physically to kill them. For example, some pertusis vaccines are prepared by chemically killing live, virulent *Bordetella pertusis* cultures. The analogies for viruses include live virus, attenuated live virus, or killed virus vaccines. Toxoid vaccines are based on a single cellular component derived from a relevant toxin. The toxins, such as diptheria toxin, are usually de-toxified by chemical treatments. Toxoids actually fall into a larger more general class of vaccines described as subunit vaccines. Subunit vaccines are prepared from subcellular components of bacteria and therefore do not contain whole cells nor all components of lysed cells. In principle, so long as the appropriate antigen(s) is selected, subunit vaccines may be prepared to protect against virtually any infectious disease. The greatest advantage of a subunit vaccine compared to bacterins is the elimination of cellular components that do not contribute to eliciting of a protective response. Some of these components may actually have inherent toxic activity, and thus their elimination is advantageous. Endotoxin is an integral part of all gram-negative bacteria, being a component of the lipopolysaccharide in their outer membranes. The biological activities of endotoxin are diverse and apparently common to endotoxins from all gram-negative species. Some of the toxic activities of endotoxin include shock, pyrogenicity, vascular permeability, and abortions in pregnant females. Although it is almost impossible to completely exclude endotoxins from products prepared from gram-negative organisms, the complications that result from endotoxemia can be so severe that it is advantageous to reduce endotoxin contamination to a minimum. This may be accomplished by preparing subunit vaccines.

Recombinant DNA techniques may be used to construct bacterial strains that are more amenable to use in bacterins or in the development of subunit vaccines. The following discussion will serve to illustrate some of the advantages of using r-DNA techniques in vaccine development. In some cases the examples are based on real knowledge of microbes and relevant antigens, while other examples represent perceived problems.

Situations where the organisms being used for vaccine production are of high virulence pose a real health threat to individuals in a production operation. Therefore, as an example, growth of *Salmonella typhi* strains, responsible for typhoid fever, would be considered prohibitive in large scale. The hazards associated with this organism and others and their potential for epidemic spread is of such a magnitude that the development of vaccines by traditional techniques would be avoided. A corollary to this problem is if the organisms were virulent and used as a live bacterin, it may very well cause the disease it was meant to control. If the mucosal surface of the small intestine was to be stimulated by live organisms to induce a local immune response, an organism that did not cause disease at the site would obviously be required. However, organisms producing the necessary antigen probably are natural pathogens. The development of a *Shigella*-based vaccine to prevent shigellosis provides an example of this problem. The intent is to induce a local immune response in the gut. However, genetically altered and attenuated strains fail to induce the appropriate antibodies, and suitable wild type strains cause disease. The growth characteristics of the organisms being used also are important. Organisms that do not grow well or that require extraordinary manipulations to promote growth are likely to be poor candidates for vaccines. One extreme of this situation pertains to viruses. The yield for many viruses is probably inadequate for large scale production and requires added time and labor when compared to bacteria. Furthermore, cultivation of viruses can be a relatively expensive operation.

Besides the inherent problems associated with the growth and virulence of some bacteria, issues of gene, and therefore antigen, expression must be considered. Antigens inappropriately displayed on a cell or produced in small amounts pose serious problems in vaccine development. Surface antigens frequently are masked as a result of cellular polysaccharide capsules and thus do not illicit good immune responses. Expression of the *E. coli* pilus antigen K99 exemplifies this phenomenon. Selection of growth conditions frequently is critical in obtaining maximal antigen expression. Growth *in vitro* often times is not appropriate for the expression of antigens produced *in vivo*. Gene instability also may pose a problem. Some genes, especially those that are plasmid encoded, are readily lost by growth *in vitro*. Thus the organism may no longer produce relevant antigen(s).

Probably the biggest problem encountered in vaccine development is determining which antigens must be included. In the case of bacterins, the process may be intuitive since killed, virulent organisms are usually used. However, in order to use gene cloning to construct appropriate vaccine strains, whether used to produce attenuated bacterins or subunit vaccines, the relevant antigens must first be identified.

The development of vaccines to prevent severe *E. coli*-induced diarrhea in neonatal livestock animals was based on the identification of several factors that directly contributed to virulence. Enterotoxigenic *E. coli* (ETEC) differ from commensal strains in at least two aspects. Firstly they produce enterotoxins that cause the small intestine to secrete water. And secondly, they have the capacity to intensively colonize the small intestines of infected animals. ETEC produce two different classes of enterotoxins: heat labile (LT) and heat stable (ST). ST subsequently has been divided into two distinct groups, STa and STb, based on solubility in methanol and spectrum of biologic activity. Both STs are poorly antigenic while LT is highly antigenic. ETEC colonize by adherence to the musocal epithelium of small intestines and subsequently proliferating at that site. Adhesion is facilitated by proteinaceous appendages on surface of the bacterial cell called pili.

Because of the protein nature of pili, they are good antigens. Antibodies produced against specific pilus-adhesions neutralize their adhesive activities and thus prevent colonization of small intestines and therefore disease.

ETEC that induce diarrhea in neonatal pigs produce one or more of three pilus-adhesins termed K88, K99, or 987P. Since all three pilus adhesins occur in nature and since one or all may be present on strains in a specific geographic location all three must be present in a vaccine designed to prevent ETEC induced diarrhea. Alternatively, prevention of symptoms by neutralizing enterotoxin activity also could be exploited in vaccine development. However, based on the known immunogenic properties of the STs this approach would likely be less useful unless the STs could be made more reactive.

The ultimate development of the *E. coli* vaccine was based on a thorough understanding of the mechanism of pathogenesis. Although this is a powerful approach, it is not always possible. Supportive information used to determine the relevant antigens to be included in vaccine may come from the results of serologic analysis. The use of serums from acute and convalescent patients has been particularly useful in determining those antigens that stimulate immune responses. Such antibodies may be protective against disease, and thus their induction would be the desired result of vaccination. Monoclonal antibodies prepared against the pathogenic organisms provide powerful tools in further defining both the important antigens as well as important antigenic determinants (epitopes) on the antigens.

To clone antigen-specifying genes it remains essential to know precisely what antigens are desired. The subsequent selection of the correct recombinant clones usually requires the use of antibodies and thus is closely inked to serologic studies.

## Methods

It is not reasonable to describe all the methods and protocols that could be used in the development of r-DNA vaccines.

### Preparation of DNA

The method used to prepare DNA for gene cloning is, in part, dependent upon the location of the specific gene. The methods employed, for example, to obtain plasmid DNA are different than those used for chromosomal DNA. Frequently, information on the gene loci are known thus facilitating the selection of methods for DNA purification. For *E. coli* pilus-adhesins, it was known that some were plasmid encoded.

If the genetic loci are not known, the procedures used to prepare chromosomal DNA should suffice since these are general methods for purifying total cellular DNA.

The procedures for cellular DNA purification can be divided into two general classes. One is based on .the procedures of Marmur which entail cell lysis, deproteinization using proteases and/or phenol extraction, and followed by selective precipitation. Various extraction and precipitation solvents have been used by various investigators. Frequently, these procedures result in material contaminated with RNA. Several enzymes used in the gene cloning process are inhibited by RNA thereby requiring an RNA-free substrate. Treatment of the DNA samples with DNase-free RNase prior to deproteinization will eliminate the vast majority of RNA. However, free ribonucleotides and oligonucleotides may co-precipitate with DNA in the presence of ethanol, thus demonstrating the need for further purification of the DNA.

The second general procedure in cellular DNA purification is based on equilibrium centrifugation. Cessium chloride is commonly used to generate salt gradients at sufficiently high concentrations such that DNA obtained, for example, by the Marmur procedure will reach an equilibrium point equal to its buoyant density. Due to the extreme differences in density of protein, RNA, and DNA these three components are readily separated resulting in highly purified DNA.

Nucleic acids also maybe separated by a variety of chromatography procedures including selective binding to and elution from hydroxyapatite.

The procedures used to purify plasmid DNA exploit two physical properties unique to these DNA molecules: their superhelical structure and their relatively small size. Gentle lysis procedures have been devised that minimized chromosome membrane d'sruption while maximizing chromosome integrity. These lysates are cleared of most of the large debris including the intact chromosome by centrifugation. The resultant supernatant contains the smaller plasmid DNA, broken, linear chromosomal DNA, RNA, and proteins. Much of the RNA and protein can be eliminated by RNase treatment followed by phenol extractions. Highly purified plasmid DNA can be ultimately prepared by separation in CsCI gradients containing the intercalating dye ethidium bromide. Alternatively the plasmids may be purified from the cleared Iysates by procedures such as reversed phase chromatography, agarose gel chromatography, agarose gel electrophoresis, or affinity chromatography (e.g., on columns of acridine tagged agarose).

The yield of plasmid from a culture is partly dependent upon size and copy number. The copy number of some plasmids may be amplified by treating the culture with certain antibiotics such as chloramphenicol or spectinomycin.

**DNA Cloning Vectors**

The intent of gene cloning is to insert the desired gene into a self-replicating DNA vector. The selection of the vector is dependent upon the source of the DNA to be cloned and the outcome desired from gene cloning. Most cloning vectors possess several common properties. Most are constructed plasmids that have readily selectable gene markers. Antibiotic resistance markers are most commonly used. Sites within the antibiotic resistance markers are used as the site of DNA cloning which results in insertional inactivation of the resistance gene. For example, the vector pBR322 encodes resistance to both ampicillin and tetracycline. Cloning into one of the resistance genes results in the inactivation of that gene. Thus cells possessing pBR322 are resistant to both antibiotics while cells containing a recombinant molecule, with the site of insertion being in the tetracycline resistance gene, are sensitive to tetracycline and resistant to ampicillin. The use of these markers and others facilitates selection of cells containing recombinant DNA molecules.

***DNA restriction endonuclease***

The capacity to insert DNA into antibiotic resistance genes is based, in part, upon the presence of specific palindromic DNA sequences, such as GGATCC, within the resistance genes. These symmetrical sequences have the same sequence on the complementary strand when read in the opposite direction. The palindromes are recognized by a class of DNases called class two DNA restriction endonucleases and thus have also been termed restriction sites. Most, if not all, bacteria produce one or more of these restriction enzymes. Their function is to restrict the entry and establishment of DNA into bacteria from genetically divergent sources. The enzymes recognize a specific palindrome cutting at a specific-site within or near the palindrome. In the absence of a repair process, such DNA molecules are lost due to their inability to be replicated in bacteria after cleavage. To prevent self-destruction, organisms produce modifying enzymes that alter certain nucelotides within the restriction site, thereby preventing cleavage by their own restriction endonucleases.

Different restriction endonucleases recognize different palindromic sequences. There are more than 200 restriction endonucleases recognizing different palindromes, many of these enzymes being available commercially. Therefore, many sites can be used in gene cloning. Typically, restriction sites within selectable genes are most useful.

Another relevant property of restriction endonucleases is that they cut DNA molecules symmetrically. If the cuts are staggered, the cleavage results in so-called sticky ends, while straight cuts result in blunt ends.

***DNA ligation***

DNA molecules from any source cleaved with the same restriction endonuclease (that results in staggered cuts) have sticky ends with the same nucleotide sequence. The sticky ends can reaneal and in the presence of the enzyme DNA ligase, a phosphodiester bond between the 3' -hydroxyl and 5' -phosphate is formed (ligation). If the sticky ends were from the same DNA molecule, a circular structure would be formed. On the other hand, if the sticky ends were from two different DNA molecules, a new hybrid molecule would result. The two ends from the hybrid can also be ligated resulting in a recombinant, circular DNA molecule. If one of the DNA molecules had been the cloning vector and the other encoding a gene, the end result would be a "cloned gene." The probability that the vector and insert DNA are ligated together rather than the ligation of several pieces of insert DNA can be increased by selection of appropriate concentrations of the various DNAs or more particularly, the concentration of 5'-ends.

Cloning of fragments cut with restriction endonucleases that create blunt-end can also be accomplished. However, the ligating enzyme must be T4 DNA ligase. It should be noted that when sticky-ended DNA is ligated, the restriction site is recreated and may subsequently be recut. Blunt-end ligations may or may not generate the original restriction sites.

***Cloning strategy***

Frequently, little is known about the genetic location of specific genes. In such cases the cloning of these genes follows an empirical process (often called "shotgun" cloning). Furthermore, in the absence of information about the number and distribution of restriction sites on and around the genes to be cloned it is difficult to design specific cloning strategies. If a restriction site is within the gene being cloned and if that enzyme is used to generate fragments to be cloned, an active gene probably will not be recovered due to its cleavage prior to cloning. To obtain an intact gene, two fragments would need to be cloned, and they must be arranged in the proper orientation. To increase the likelihood of cloning genes, several techniques have been employed. Each makes use of randomizing the locations of restriction cuts. This may result in the generation of large fragments.

Obviously, the larger the fragment, the greater the probability of cloning a gene and subsequently identifying such clones from a large DNA pool. The randomizing of cut sites decreases the chances that a cut will occur within the gene. Controlled shearing of DNA will result in fragments with a given size range. Partial digestion with restriction endonucleases can also be employed to obtain large fragments. In this regard, enzymes that recognize 4-base pair palindromes, such as Sau 3AI, are particularly useful since such sites, if randomly distributed, would occur every 256 base pairs. Partial digests with Sau 3AI should result in a random distribution of cuts and therefore fragments. In the case of sheared DNA, homopolymer tailing of the fragments and of the cloning vector is necessary. Otherwise, recognition sites for annealing would not be available, resulting in an inability to ligate and thus clone the fragment. Alternately, the ligation of linkers (synthetic DNA sequences containing a restriction site) to fragments may be performed and used as a means for gene cloning. For example, Bam HI linkers may be blunt end ligated to fragments generated by shearing. If the shearing resulted in single stranded regions at the termini, it would be necessary to make them blunt ended prior to ligation by either removing the single stranded region using nuclease S1 or by filling in the single stranded region with DNA polymerase. The fragments could then be inserted and ligated into a Bam HI site in a vector.

In generating random cuts, it often is difficult to restrict the fragments to a narrow size range. The amount of fragments of certain sizes may be increased by separation and elution from molecular sieves or by electrophoresis through agarose.

Recombinant DNA molecules are usually introduced into bacteria by transformation. Most of the best understood cloning systems were developed in *E. coli* and therefore transformation most frequently is into *E. coli*. The procedure of Cohen et al. and its many derivations are used for transformation. Efficiencies of transformation are dependent upon the host strains and the size of the DNA molecules. Several *E. coli* strains have been selected that are transformed more efficiently than other strains. Furthermore, the transforming DNA must be in a covalently closed circular configuration for efficient uptake. Of major consideration in selecting recipient strains are the effects of host restriction and modification enzymes on the incoming DNA. It is certain that host restriction enzymes if functioning properly would eliminate the incoming DNA unless it had been derived from a strain with a functional modification system. Restriction and modification mutants are usually employed for transformation to eliminate this possibility. If significant DNA sequence homology exists between the host and incoming DNA, the risk of homologous recombination can be reduced using *recA* mutants.

***Cosmid cloning***

As the size of the DNA to be transformed into a strain increases, the efficiency of the process decreases. For the situations described above where fragment sizes are kept large, transformation frequencies are often low. In many cases, the frequency is too low to allow convenient cloning. In these situations, the use of cosmids may solve the problem.

Cosmid cloning vectors are constructed by combining part of a conventional plasmid cloning vector with a bacteriophage lambda sequence called *cos*. *Cos* (or cohesive ends) is a sequence necessary to package DNA into lambda. The procedure proceeds by ligating a fragment into a cosmid cloning vector as usual except that the DNA concentration is kept high to promote concantamer formation. The ligated DNA is then packaged *in vitro* into lambda heads and tails and transduced into a lambda sensitive strain. Since the only lambda sequence in the cosmid cloning vector is *cos*, the only way the DNA can replicate is to use the plasmid-specified functions (also encoded in the cosmid cloning vector). In order to be packaged into lambda heads and tails DNA must, in addition to having *cos*, be approximately 45 kilobases in length. This means that 30 to 40 kb of insert DNA may be cloned. Conventional plasmid cloning due to size limitations in transformation is usually restricted to inserts of approximately 10-15 kb.

The selection of the cloning vector to be used is ultimately dependent upon the DNA source and expected outcome of cloning. A somewhat empirical approach must be used. Several considerations include: what enzymes will be used and the size of the fragments generated; will the cloned DNA be expressed in *E. coli*; will the end product of gene expression be degraded by the host; if it is an intergeneric cloning experiment, will the hybrid DNA molecule be expressed in the other genus? Several expression vectors have been constructed that contain appropriate promoter sequences. Expression of genes in intergeneric cloning may require introduction back to the "natural" host. Shuttle vectors that can replicate in different genera have been constructed for this purpose. The prevention of gene product break down may be prevented in some cases, by fusion of the protein to another of host origin. Cloning vectors that perform this function also are available.

**Identification of Recombinants**

The success of gene cloning is also dependent upon the ability to detect cells containing the gene. Detection of the desired recombinants have been based on two properties. The first and most desirable property is the synthesis of the desired gene product. Gene products that have enzymatic activity may be detected using an appropriate enzyme assay. However, most gene products that would be used in vaccines are not enzymes and must be detected as antigens. Antigens present on the cell surface may be detected by serum agglutination either in suspension or by precipitation reactions in agar. Colonies also may be transferred to sheets of polyvinyl or nitrocellulose, left intact or lysed, and subjected enzyme linked immunoassays or radioimmunoassays. If DNA probes are available, recombinants can be detected by colony hybridization. One of the most sophisticated approaches is based on the synthesis, in the laboratory, of specific DNA probes. The product of the gene to be cloned is purified and a partial amino acid sequence determined. Based on the amino acid sequence, short oligonucleotides are synthesized that represent the possible coding sequences. In a bacterial system a 10-12 base pair probe would be sufficiently large to eliminate chance that the same sequence would occur in another gene. The DNA probes would then be used in colony hybridization procedures to detect the putative recombinants.

## Development of an r-DNA Vaccine to Prevent ETEC Induced Diarrhea

**Rationale for Development**

As mentioned above, the pathogenesis of ETEC induced diarrhea is depenent upon the expression of two known attributes of virulence: 1. secretion of enterotoxin that induces net water efflux by the small intestine; and 2. colonization of small intestines by adherence to the mucosal surfaces. Adherence of ETEC to small intestines is mediated by protein, surface appendages called pili. In baby pigs, three antigenically different pili are known to facilitate adhesion: K88, K99, 987P.

Initially, it was shown that baby pigs suckling dams that had been vaccinated individually with purified pili were protected against severe, fatal diarrhea after experimental challenge with ETEC producing the same pili as were in the vaccine but not from ETEC producing different pili. The apparent mechanism of protection was believed to be due to the consumption of colostrum containing pilus-specific antibodies. The antibodies neutralized the adhesive activity of the pili, thus preventing the ETEC strain from adhering and colonizing the small intestine. Some evidence indicated that these antibodies also could reverse the adhesion of already adherent bacteria.

**Table 12.1. Affect of vaccination with 987P on mortality of pigs challenged with ETEC producing 987P or K99.**

| *Vaccine* | *Pilus on challenge ETEC* | *Number of pigs challenged* | *Number that died (%)* |
|---|---|---|---|
| 987P | 987P | 75 | 1 (1%) |
| | K99 | 30 | 9 (30%) |
| K99 | 987P | 41 | 0 (0%) |
| | K99 | 80 | 7.2 (17%) |
| | 987P | 35 | 14 (40%) |
| Control | K99 | 78 | 12.5 (22%) |

The next step was to determine whether a vaccine containing all three pilus-types would have a broad-spectrum, protective effect. The results shown in Table 12.2 demonstrate that vaccination with a cell-free mixture of the three pili protected against disease after the pigs were challenged with ETEC producing K88, K99, or 987P.

**Table 12.2. Effect of vaccination with EcoBac on mortality of challenged animals.**

| *Vaccine* | *Pilus on challenge ETEC* | *Number of pigs challenged* | *Number that died (%)* |
|---|---|---|---|
| | K88 | 32 | 5 (15.6%) |
| | K99 | 34 | 1 (2.9%) |
| EcoBac | 987P | 44 | 0 (0%) |
| | K88 | 47 | 29 (61.7%) |
| Control | K99 | 48 | 27 (56.2%) |
| | 987P | 44 | 31 (70.4%) |

The rationale behind usage of r-DNA techniques for the development of this vaccine was based primarily on two potential problems: 1. the K88 plasmid was readily lost by cultivation *in vitro*; 2. the yield of

K99, especially compared to K88 and 987P, was very low. Both problems were solved by gene cloning.

**Cloning of *E. coli* Pilus Genes**

The cloning strategy for two of the three *E. coli* pilus-specifying genes used to produce Eco Bac, a vaccine to prevent ETEC diarrhea in pigs, was based on the observation that these genes were plasmid encoded.

The K99 specifying plasmid (pK99) from a virulent, wild type ETEC strain was moved into an *E. coli* K 12 strain by conjugal mating. The plasmid was subsequently isolated from a 500 ml culture grown in trypticase soy broth at 37°C. A lysate was prepared using the procedure of Ish-Horowicz and Burke. Cells were harvested at 10,000 × g and resuspended in 0.1 volumes glucose (0.1 M), EDTA (0.01 M), tris (0.05M, pH 8). After five minutes at room temperature, two volumes of SDS (1%) in 0.2N NaOH were added to lyse the cells and denature the DNA. After an additional five minutes on ice, 1.5 vols potassium acetate (5M, pH 5) was added and incubated on ice five minutes. The resulting precipitate, including membranes, SDS and chromosome, was removed by centrifugation at 10,000 × g for ten minutes. To the supernatant, 2.5 volumes ice cold ethanol (95%) were added and the mixture stored at least 30 minutes at -20°C. The precipitate was collected, dried in vacu, and resuspended in 10.6 ml Tris (0.01 pH 8), EDTA (1 mM, TE). To the dissolved DNA, 2.1 ml ethidium bromide (1 mg/ml) and 8 gms CsCl were added and the mixture centrifuged to equilibrium. The RNA, covalently closed circular plasmid DNA, linear and open circular DNA (chromosome fragments and plasmids), and proteins and membrane debris separate into four different regions of the gradient based on their bouyant densities in CsCl. The CCC plasmid DNA was removed, extracted with butanol to remove ethidium bromide, ethanol precipitated and dried.

Vector DNA (pBR322) was prepared in the same manner, except that chloramphenicol (150 μg/ml) was added to the log phase cells to amplify plasmid synthesis.

Both vector and pK99 DNA were cleaved with the restriction endonuclease BamHI using the conditions described by the manufacturer. The 5' -phosphate group of pBR322 was removed by addition of calf intestine alkaline phosphatase. Both the digested and diphosphorylated vector and digested pK99 were phenol extracted (two times), ether extracted (three times), and precipitated with ethanol. The precipitates were collected, dried, and resuspended in TE. One $\mu$g of pBR322 and 0.5 g pK99 digest were mixed together. Two $\mu$l of ligation buffer (0.66 M tris, pH 7.5, 50 mM $MgCl_2$ 50 mM dithiothreitol and 10 mM ATP) were added and the mixture adjusted to 20 $\mu$l. T4 DNA ligase was added (two units) and the mixture incubated 18 hours at 10°C. One half of the mixture was used to transform competent *E. coli* cells prepared by the procedure of Cohen, et al. Cells were plated on trypticase-soy agar containing ampicillin (25 $\mu$g/ml). After overnight incubation, colonies were replica plated to ampicillin plates and tetracycline (15 $\mu$g/ml) plates. Colonies containing ampicillin resistant tetracycline sensitive cells were screened for K99 production by serum agglutination using rabbit antisera prepared against purified K99.

Cloning of K88 genes was identical except that the DNA source (pK88) and antiserum (antiK88) were different, and the restriction enzyme employed was Hind III instead of Barn HI.

Identification of the most useful clones (i.e., most stable and best expression), quantitation and the ultimate formulations for the final vaccine were empirical.

**Subcloning**

It often is desirable to subclone DNA fragments particularly when the fragment is large as is the case in cosmid cloning. This can be performed using different restriction enzymes or by varying the length of time of digestion (for recombinants derived from partial digests).

For most practical applications of gene cloning, the subcloning process is to remove unnecessary DNA sequences. Frequently, recombinant DNA molecules are unstable, this phenomenon often being associated with the size of insert. Thus plasmids constructed with large inserts may be less stable than those with small inserts demonstrating a need to subclone. Unfortunately, this is not a hard and fast rule. Other properties of the DNA molecules (other than size) are important in determining stability. The hybrid K99 molecule described above is 11.15 kb in size. Upon successive subculturing (three passages), the plasmid was lost at high frequency. This problem can be circumvented, however, by selectively growing cells to maintain the plasmid. For the example of the K99 recombinant, the culture medium is supplemented with ampicillin. Since the K99 recombinant DNA molecule encodes ampicillin resistance (a vector encoded trait), only those cells containing the plasmid grow. Loss of inserted DNA has not been a problem for either K99 or K88 specifying recombinant DNA molecules.

## Final Remarks

The ease of development of the *E. coli* vaccine was greatly enhanced since the genes being cloned were from *E. coli* and were to be expressed in *E. coli*. Furthermore, since the gene products being exploited (pilus-adhesins) are surface antigens, their transport to the cell surface in the constructed recombinant strains occurred in an unaltered manner.

Several problems envisaged in construction of strains for bacterial vaccines relate to issues of the cellular location of the antigens and whether they are even expressed in the constructed strains. For example, genes from eucaryotic sources probably will not be expressed in procaryotes. Likewise, animal virus genes probably would not be expressed in procaryotes. However, gene cloning is most readily performed in bacteria and particularly in *E. coli*. Thus, shuttle vectors must be employed. However, there are few shuttle 'vectors available, and therefore the number of eucaryotic hosts (or for that matter other bacterial hosts) that can be used is equally low.

*Bacillus subtilis* has been used as a procaryotic host for gene cloning. The particular advantage of this organism over *E. coli* is that some proteins are readily secreted simplifying recovery and purification of the proteins in commercial applications. However, the number and diversity of vectors for *B. subtilis* is low greatly reducing its current usefulness in generalized commercial applications. Furthermore, *B. subtilis* produces potent proteases that may degrade the desired material, if composed of protein prior to its isolation.

The recent development of methods for viral gene cloning and expression in vaccinia virus has greatly facilitated the developments of vaccines for viral diseases. As the techniques of gene cloning are improved and refined and the techniques to enhance gene expression and antigen transport are developed, the overall progress in r-DNA vaccine development will accelerate as will all areas of biotechnology using r-DNA techniques.

# Index


A

A. glaucus, 116

A. hydrophila, 90

A. parasiticus, 79

Acetobacter, 21, 22, 63

Acetobactor (Mycoderma) aceti, 22

Adhatoda vasica, 233

Adjuvants, 15

Aeration, 8

Aerobic, 22

Aerobic, 30

Agrobacterium tumefaciens, 206, 222

Alkaligenes flavobacterium, 31

Anaerobically, 31

Anisodus tanguticus, 233

Artemia salina, 222

Artemisia annua, 235

Ashbya gossypii, 27

Aspergillus, 29, 32

Aspergillus glaucus, 116

Aspergillus nigar, 78, 86

Atropa belladonna, 227

Autographa californica, 210

B

B. cereus, 75

B. coagulans, 105, 109

B. fulva, 86

B. megaterium, 102

B. subtilis, 82, 105, 285

B. volutans, 19

Bacillus, 23, 27, 31, 32, 79, 107, 160

Bacillus licheniforrnis, 109

Bacillus subtilis, 24, 285

Bacillus thuringiensis, 209

Batch, 3, 7

Beauveria bassiana, 208

Beer, 9, 10

Benomyl, 84

Berberis vulgaris, 232

Beta rays, 131

Biochemical oxygen demand, 34

Bioconversion, 2, 29

Biphenyl, 84

Bombyx mori, 211

Bordetella pertusis, 272

Botrytis, 64, 84, 88

Browning, 268

Burning, 59

Butylated hydroxyanisole, 74

Butylated hydroxytoluene, 74

Byssochlamys fulva, 86

**C**

C. acutifolia, 233
C. albicans, 73
C. botulinum, 62, 63, 66, 67, 68, 69, 70, 71, 74, 75, 78, 79, 80, 81, 84, 90, 91, 101, 110, 111
C. ledgeriana, 231
C. perfringens, 66, 70, 71, 117
C. purpureumis, 205
C. roseus, 230
C. sporogenes, 66, 70, 80
C. tyrobutyricum, 66
Camptotheca acuminata, 235
Candida, 64, 88, 139
Candida albicans, 73
Candida krusei, 135
Candidae shihatae, 56
Cassia angustifolia, 233
Catharanthus roseus, 229
Cathode rays, 131
Cellulomonas, 31
Cephaeline, 232
Cephaelis ipecahuanha, 232
Champagne, 13
Chemical oxygen demand, 34
Chemostat, 4
Chondodendron tomentosum, 220, 228
Cinchona, 231
Cinchona ledgeriana, 225
Citric, 84
Cl. acetobutylicum, 19, 20
Cl. butylicum, 20
Cl. felsineum, 19
Cladosporium, 88
Clostridium, 19, 31, 32, 66, 79, 107
Clostridium botulinum, 135
Clostridium butyricum, 79
Container, 268
Continuous, 3
Corynebacterium, 29
Coxiella burnetti, 100, 109
Cracking, 8
Cryptosporidium parvum, 231
Cultivation, 1
Curvularia, 29
Cytisus scoparius, 233
Cytophaga, 31

**D**

D. lanata, 223
D. purpurea, 223
Darkening, 269
Datura metel, 227
Datura stramonium, 227
Debaryomyces, 88
Dehydroacetic acid, 86
Desulfovibrio, 190, 191, 192
Desulphovibrio, 32
Diffuser, 8
Digitalis, 223
Dioscorea, 220
Distillation, 16

**E**

E. coli, 15, 56, 73, 78, 142, 143, 144, 145, 147, 150, 151, 153, 154, 160, 164, 168, 171, 274, 275, 279, 280, 283, 284, 285
Electrons, 131
Electronvolt, 131
Ellipsoideus, 12
Endoenzymes, 24
Entamoeba histolytica, 232
Entering, 268
Enterococcus (Streptococcus) faecalis, 79
Ephedra sinica, 229
Erwina chrysanthemi, 56
Erwinia, 206
Erythroxylon coca, 225
Escherichia coli, 56, 129, 142, 272

Ethanol, 86
Exoenzymes, 24

F

Fermenter, 4
Flavobacterium, 88
Flavobacterium proteus, 15
Fluidized bed biofilm reactor, 39
Fusarium, 84

G

Gamma rays, 131
Geotrichum, 88

H

H. zea, 210
Hafinia alvei, 63
Halodurics, 62
Halophiles, 62
Harvesting, 2
Heliothis, 210
High fructose corn syrup, 174
High wine, 9
Homeostasts, 7
Hydraulic retention times, 39
Hydrogen peroxide, 85
Hydroxyl, 83
Hypertonic, 62

I

Impeller, 8
In vitro, 140, 144, 148, 152, 153, 154, 155, 274, 280, 282
In vivo, 153, 274
Inter alia, 212
Intermediate-moisture foods, 119

K

K. oxytoca, 56
Kilorad, 131
Klebsiella oxytoca, 56
Klebsiella planticola, 56

L

L. arabinosus, 30
L. buchneri, 21
L. Dextranicum, 31
L. dispar, 216
L. plantarum, 29, 68
Lactobacillus, 8, 12, 19, 31, 100
Lactobacillus casei, 30
Lactobacillus delbrueckii, 20
Lactobacillus lactis, 30
Lactobacillus pastorianus, 15
Lactobacillus plantarum, 29, 30, 86
Lactoperoxidase system, 87
Leptinostarsa decemlineata, 209
Leuconostoc, 31
Leuconostoc mesenteroides, 31, 86
Liquid, 251
Liquor, 16
Listeria monocytogenes, 90
Lymantria dispar, 216

M

Malt, 19
Malting, 13
Mash, 6, 8
Mashing, 15
Megarad, 131
Megarep, 131
Metabolic byproduct, 2
Methanobacterium bryantii, 189, 190
Methanobacterium soehngenii, 189
Methanobacterium thermoautotrophicum, 193
Methanosarcina barkeri, 196
Methanosarcinae, 189, 195, 196
Methanospirillum hungatei, 192
Methanothrix soehngenii, 189, 196
Metmyoglobin, 65
Microbacterium, 135
Microbiological assay, 30
Micrococcus, 31
Micrococcus roseus, 135
Minimal inhibitory concentrations, 73
Mother-of-vinegar, 22

Mucor, 88
Mycobacterium hominis, 109
Mycobacterium tuberculosis, 100
Mycobaterium, 32
Mycoderma aceti, 21

N

Neurospora, 30
Nitrosamines, 68
Nitrosomyoglobin, 65
Nocardia, 32
Nosema locustae, 208

O

Optimum, 5
Osmoduric, 63
Osmophiles, 63
Oxymyoglobin, 65

P

P. aeruginosa, 75
P. falciparum, 231, 236
P. fluorescens, 73
P. geniculata, 73
P. vulgaris, 61
Pachysolen tannophilus, 56
Packaged liquors, 10
Papaver somniferum, 226, 228
Pediococcus acidilactici, 68
Pediococcus cerevisiae, 15
Penicillium, 88
Penicillium chrysogenum, 4
Penicillium notatum, 7
Peppery, 59
Physostigma venemosum, 227
Pichia stipitis, 56
Picowaved, 132
Pilocarpus jaborandi, 227
Piper mythesticum, 225
Plantarum, 68
Plasmodium falciparum, 231, 235
Plasmolysis, 62
Pneumocystis carinii, 231
Podophyllum peltatum, 230
Precursors, 23
Productive mutants, 4
Propionibacterium, 27
Proteus, 31
Proteus vulgaris, 61
Pseudomonas, 31, 32, 88, 206
Psychrophiles, 88
Pure, 8
Pure culture, 7
Pyridoxine, 30

R

R. serpentina, 223, 224
R. vomitoria, 224, 225
Radappertization, 132
Radicidation, 132
Radurization, 132
Rauwolfia, 220, 224
Rauwolfia serpentina, 223
Rhizopus, 29
Rhodotorula, 88
Roentgen, 131
Roentgen-equivalent-physical, 131

S

S. aureus, 66, 73, 75, 78, 96, 109, 117, 124
S. bayanus, 64
S. carlsbergensis, 15, 64, 139
S. cerevisiae, 15, 56, 139, 148
S. faecalis, 68, 73, 117
S. faecium, 68
S. lactis, 68
S. marcescens, 63
S. senftenberg, 102, 105, 106
S. typhimurium, 75
S. uvarum, 139
Saccharomyces, 15, 16, 64, 139
Saccharomyces cerevisiae, 8, 12, 21, 23, 27, 56, 86

Saccharomyces rouxii, 63, 116
Salmonella senftenberg, 102
Salmonella typhi, 273
Sanguinaria canadensis, 233
Serratia, 31
Serratia liquefaciens, 63
Shigella, 232, 273
Silent Spring, 204
Sodium diacetate, 84
Sparkling wines, 13
Spodoptera exigua, 210
Staphylococcus aureus, 28
Steady state, 7
Still wines, 12
Str. liquefaciens, 32
Streptococcus, 100
Streptococcus lactic, 79
Streptomyces, 27, 29
Streptomyces natalensis, 81
Strophanthus gratus, 225
Submerged cultures, 7
Sulphite liquor, 22
Syntrophobacter wolfei, 192
Syntrophobacter wolinii, 192
Syntrophomonas wolfei, 192

T

T. reesei, 52, 53, 54
Taxus brevifolia, 234
Temperature coefficient, 88
Thiamine, 30
Thibendazole, 84
Titratable acidity, 76
Torula, 22
Torula utilis, 23, 27
Torulopsis, 88
Torulopsis candida, 78
Torulopsis utilis, 22
Trichinella spiralis, 98, 117
Trichoderma, 52
Trichoderma reesei, 53
Trichogramma, 208
Two-phase continuous process, 22

U

Upflow anaerobic sludge blanket, 37

V

V. parahaemolyticus, 78, 90
Vapour, 251
Very enriched fructose corn syrup, 184
Vinca rosea, 229
Vitis, 10
Vitis vinifera, 10

W

Wine, 10
Wood smoke, 86
Wort, 15

X

Xanthomonas, 206
Xeromyces bisporus, 116
Xylose, 22

Y

Y. enterocolitica, 90

Z

Z. mobilis, 56
Zygosaccharomyces bailiff, 64
Zymomonas mobilis, 56